Machine Learning Plasmas and the Neuromorphic Plasma Chemistry

Machine Learning Plasmas and the Neuromorphic Plasma Chemistry

edited by

Li Lin

Michael Keidar

JENNY STANFORD
PUBLISHING

Published by

Jenny Stanford Publishing Pte. Ltd.
Level 34, Centennial Tower
3 Temasek Avenue
Singapore 039190

Email: editorial@jennystanford.com
Web: www.jennystanford.com

British Library Cataloguing-in-Publication Data
A catalogue record for this book is available from the British Library.

Machine Learning Plasmas and the Neuromorphic Plasma Chemistry

ISBN 978-981-5129-84-7 (Hardcover)
ISBN 978-1-003-71013-4 (eBook)
DOI 10.1201/9781003710134

Contents

Preface

Plasma engineering has long stood as a vital pillar of modern science and technology, shaping advances in energy, manufacturing, space propulsion, and, more recently, medicine. In recent years, a new synergy has emerged at the intersection of plasma science and artificial intelligence: machine learning (ML) is proving to be a powerful tool for both understanding and manipulating the complex dynamics of plasmas. This book explores this exciting confluence and looks forward to an even more radical possibility—the emergence of intelligent matter.

This book is written for students, researchers, and engineers at the undergraduate level and above who are eager to explore how machine learning is being used to enhance plasma engineering today and how it may reshape the field in the future. While the book is not intended to be a comprehensive research volume, it is rich with conceptual depth and interdisciplinary insights that aim to inspire the next generation of plasma and AI scientists. It is written from the perspective of a plasma engineer navigating the integration of machine learning into complex physical systems. Rather than aiming to advance machine learning theory itself, the focus is on identifying domain-relevant opportunities, constraints, and innovations that emerge when ML meets real plasma problems. Readers should not expect a generic ML textbook, but rather a practical guide that bridges disciplines while staying grounded in plasma science.

Chapter 1 introduces the reader to the landscape of modern plasma engineering, surveying key application areas such as low-temperature plasma medicine, electric propulsion, and plasma-based materials processing techniques like etching and plasma-enhanced chemical vapor deposition (PECVD). These applications serve as an entry point to discuss how machine learning is beginning to assist and accelerate innovation in each of these areas. We also introduce the idea of plasma chemical pathway networks (CPNs) in Chapter 1. Unlike traditional Chemical

Reaction Network Theory (CRNT), where nodes represent reaction complexes and the focus lies in analyzing system stability and dynamic behavior, the CPN takes individual chemical species as nodes, rather than complexes. The connections represent reaction pathways that transform one species into another, analogous to information flow in neural networks which will be discussed in Chapter 4.

Chapter 2 transitions into the role of machine learning in plasma diagnostics and control. It provides foundational knowledge of neural networks and explains how machine learning models can be integrated into closed-loop control systems, using feedback from plasma and system diagnostics to guide plasma behavior in real time. This practical, systems-level viewpoint is essential for understanding how ML can improve plasma operation stability, efficiency, and precision.

Chapter 3 revisits the fundamentals of plasma science, focusing on the theory of collisions and how they govern the rich chemistry of plasmas. This background provides the essential groundwork for Chapter 4—the most innovative and forward-looking chapter of the book.

Chapter 4 presents a new paradigm: the concept of intelligent plasma. Drawing a mathematical analogy between plasma CPNs and artificial neural networks, we propose that plasmas themselves, via their internal reaction networks, can be trained and utilized as programmable, intelligent materials. In this framework, species densities act as neurons, while reaction rate coefficients behave similarly to activation functions. The chapter compares them with conventional neural network architectures and explores how control parameters like temperature can serve as the weights in CPNs. Different from CRNT studying the dynamics and equilibrium of a system, the CPN framework opens the door to the radical idea of material-based artificial intelligence: an intelligence embodied not in computer algorithms, but in matter itself. We thus suggest that this could represent a new phase of matter: an intelligent phase.

This book would not have been possible without the mentorship, collaboration, and support of many esteemed colleagues. I would like to thank Prof. Taeyoung Lee for his invaluable mentorship in machine learning and control theory, which laid the foundation for many of the ideas presented here. I am also deeply grateful to Dr. Yevgeny Raitses and Dr. Sophia Gershman at the Princeton Plasma Physics Laboratory, and to Dr. Jonathan Frank and Dr. Sebastian Pfaff at Sandia National Laboratories, for their support of my research on physics-informed neural networks (PINNs), conducted under the support of the U. S. Department of Energy.

We hope that this book will not only introduce readers to the current applications of ML in plasma engineering but also inspire new interdisciplinary approaches and bold thinking about the future of intelligent systems and programmable intelligent matter.

Li Lin
Michael Keidar
Washington, DC
July, 2025

Chapter 1

Modern Plasma Engineering and Applications

Li Lin, Anmol Taploo, Vikas Soni, Alex Horkowitz, and Carles Corbella Roca

School of Engineering and Applied Science, The George Washington University, Washington, DC 20052, USA

lilin@email.gwu.edu

1.1 Introducing Fundamental Plasma Physics

Plasma is one of the states of matter in addition to solid, liquid, and gas. A strict definition of plasma is a matter that meets all the following requirements: (1) it is quasi-neutral with charged particles, (2) it contains massive number of particles so that they behave statistically, (3) there are non-neglectable Coulomb collisions among particles, so the plasma cannot be too sparse and with significant number of charges. Therefore, the following examples are not plasmas: ideal gases (not enough charged particles with significant Coulomb collisions), a pair of positive and negative ions (not enough number of particles to use statistical method to study their physics). The portion of ionized particles over

Machine Learning Plasmas and the Neuromorphic Plasma Chemistry
Edited by Li Lin and Michael Keidar

ISBN 978-981-5129-84-7 (Hardcover), 978-1-003-71013-4 (eBook)
www.jennystanford.com

DOI: 10.1201/9781003710134-1

the total number of particles is thus a useful parameter, namely "ionization degree". In addition, plasma can also be considered as a type of magnetohydrodynamics (MHD) fluid, with complex relations between the plasma (especially plasma waves) and the external electromagnetic fields. But in addition to the complicated wave physics in plasma, plasma chemistry is also a sophisticated topic.

Both man-made and natural plasmas are common, such as lightening, sun, electric arc, nuclear fusion, etc. One can generate plasma from other states of matter, such as gases. The general rule is to apply energy on a matter so that atoms and molecules can receive energies higher than their first ionization threshold and be ionized. The most straightforward example is to heat gases. When the temperature is increased, the kinetic energy of thermal motion is increased. It can be computed, assuming there is only one type of species in the system, as

$$\frac{1}{2}mv_{\text{th}}^2 = \frac{f}{3}k_B T \tag{1.1}$$

where m is the mass of a particle, v_{th} is the average thermal velocity, f is the degree of freedom (DOF), k_B is the Boltzmann constant, and T is the bulk temperature. The left-hand side is obviously the kinetic energy of particles. There are 3 translational DOFs for monatomic molecules and 2 additional DOFs, vibrational and rotational, for diatomic molecules. When there is an elastic collision, the particles that are involved have both the conservations of energy and momentum. But there is also a possibility of inelastic collision that a part of kinetic energy of the relative velocity of them before the collision is applied on the electron cloud of the particles. This results in chemical reactions, such as breaking of chemical bonds or release of internal electrons (ionization). Figure 1.1 summarizes the commonly seen collision types with examples in plasmas.[1–5]

In Fig. 1.1(A), 9 collisions are included: elastic collision, attachment, detachment, excitation, photon emission, quenching, ionization, recombination, and charge transfer. Each type of collision can also be associative (forming new chemical bond), dissociative (breaking existing bond), or no bond change. In Fig. 1.1(A), the 27 conditions are all shown with examples except the elastic collision that cannot have any bond changes. Attachment refers to the process where an electron attaches to a neutral atom or molecule, forming a negative ion. For example, in the attachment reaction e + $O_2 \Rightarrow O^- + O$, an electron attaches to an oxygen molecule, producing a negative ion and a neutral oxygen atom. Conversely, detachment occurs when a negative ion releases an electron, becoming neutral.

(A)

	Associative	**No Bond Change**	**Dissociative**
Elastic Collision	N/A	A + B => B + A	N/A
Attachment	e+2H => H_2^-	e+2O_2 => O_2^- + O_2	e+O_2 => O^- + O
Detachment	H_2 + O^- => H_2O + e	NO + NO^- => 2NO + e	O_3 + O_3^- => 3O_2 + e
Excitation	2H => $H_2(B^1\Sigma_u^+)$	e + He => He(2^3S) + e	e + CO_2^+ => CO($A^1\Pi$) + O
Photon Emission	2H => H_2 + hv(121.57nm)	$N_2(C^3\Pi_u$ v'=0) => $N_2(B^3\Pi_g$ v"=0) + hv(337.13nm)	H_2^+ + e => 2H + hv(121.57nm)
Quenching	O(3^5S^o) + CO => CO_2	OH($A^2\Sigma^+$ v'=0) + M => OH + M	OH($A^2\Sigma^+$ v'=0) + M => O + H + M
Ionization	He + He(3^1S) => He_2^+ + e	e + H_2 => H_2^+ + 2e	e + H_2O => H^+ + OH + 2e
Recombination	C^+ + CN^- => C_2N + hv	e + He^+ => He	e + H_2O^+ => OH + H
Charge Transfer	H_2 + OH^+ => H_2O^+ + H	H^+ + O => O^+ + H	O_2^+ + N_2O_5 => NO_2^+ + NO_3 + O_2

(B)

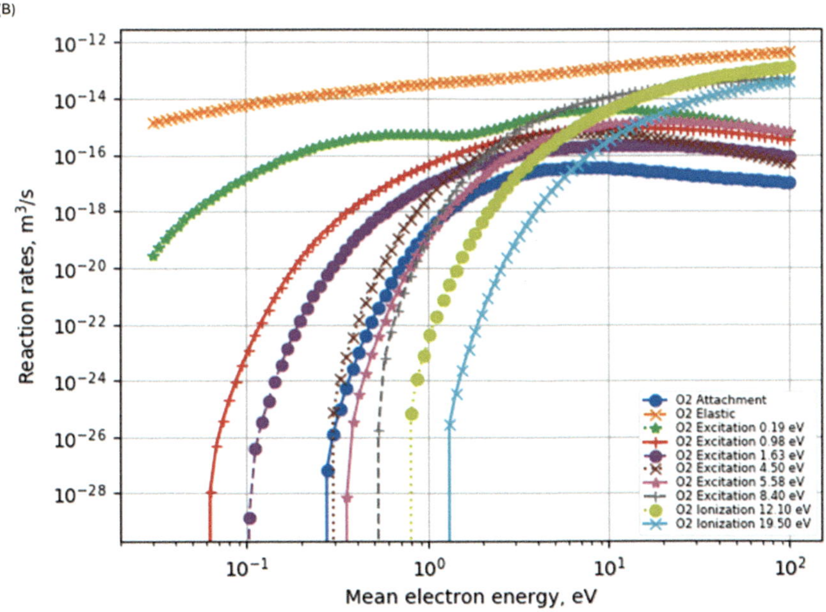

Figure 1.1 Examples of collisions in plasmas.[1–5] (A) A table summarizing the types of collisions. (B) The examples of rate coefficients of electron colliding with O_2 molecules that are acquired using the Boltzmann solver BOLSIG+.

This can happen spontaneously or through interactions with other particles, leading to the regeneration of free electrons in the plasma. Excitation occurs when a particle, typically an electron, collides with an atom or molecule, transferring energy to the latter and promoting it to an excited state. In Fig. 1.1(A), $e + He \Rightarrow He(2^3S) + e$ represents the excitation of a helium atom by an electron to the triplet metastable state "2^3S". Each excited state has a lifetime and eventually will drop to a lower state by emitting a photon if there is no other collision. The unique photon wavelength corresponds to the energy difference between the two excited states and is used in many spectroscopies to identify the chemical species. Quenching is the process by which an excited species returns to its ground state, either by emitting radiation (spontaneous emission) or through collisional de-excitation, where the excess energy is transferred to another particle. Ionization is the process by which an electron is removed from an atom or molecule, resulting in the formation of an ion. This can occur through direct collisions with energetic electrons, as in $e + H_2O \Rightarrow H^+ + OH + 2e$, where a water molecule is ionized to produce a proton, a hydroxyl radical, and two electrons. Recombination, on the other hand, involves the capture of an electron by a positive ion, leading to the formation of a neutral atom or molecule. This process is critical in reducing the ion density and restoring neutral species in the plasma. Charge transfer involves the transfer of an electron from one ion or neutral species to another. This process can occur through collisions between ions and neutral atoms or molecules. For instance, in Fig. 1.1(A), in the plasma containing hydrogen and oxygen, a reaction like $H^+ + O \Rightarrow O^+ + H$ may occur, where a proton transfers its charge to an atomic oxygen by taking its electron, resulting in the formation of a positive ion of atomic oxygen and a neutral hydrogen atom. Charge transfer is essential in shaping the ion composition and electrical conductivity of plasma.

Considering that the chemical species can be generated and consumed during their collisions, the chain of generation and consumption is called the chemical pathway to a species, such as $e + O_2 \Rightarrow O^- + O$, and $O + H \Rightarrow OH$. Since many chemical species can occur at the same time, a map of all the possible pathways can be plotted to visualize a network of pathways, namely chemical pathway network (CPN) as shown in Fig. 1.2.

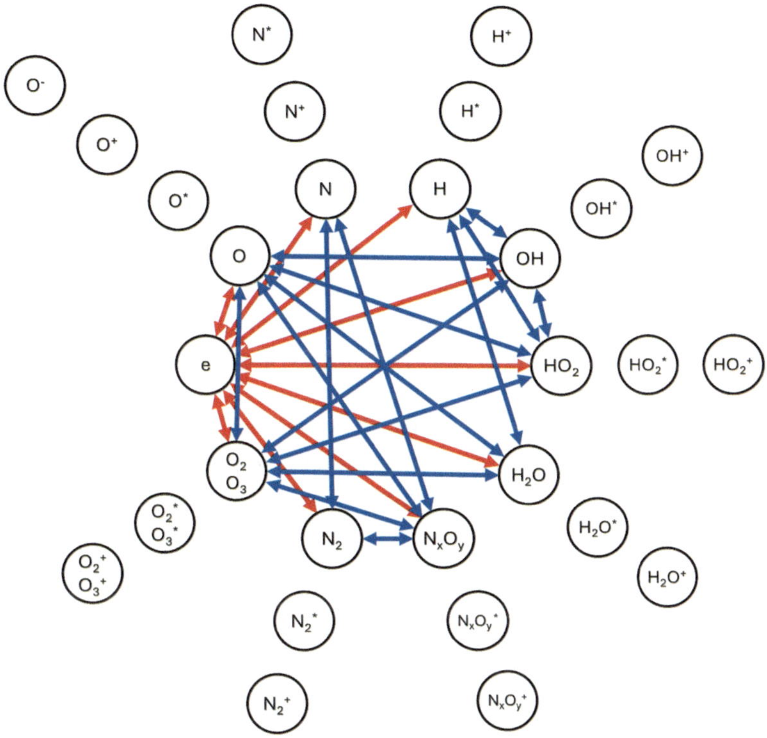

Figure 1.2 An example of CPN of air plasma where each atomic and molecular species can also have their excited states and ion form also shown in the network but not connected to make a clearer view of the network. Please note that the photons with different wavelengths are also considered as different species which will be discussed in a later chapter, but they are not shown in this figure.

The interaction between particles is often characterized by parameters such as collision kinetic energy, collision cross section, and collision rate coefficient. The collision kinetic energy E_k is the kinetic energy of the relative velocity of the colliding particles. The collision cross section σ quantifies the effective area over which collisions occur, representing the probability of a collision given a certain relative velocity. The cross section can depend on the kinetic energy of the particles, as different interaction mechanisms (such as elastic scattering, ionization, or excitation) have varying probabilities at different energies. The collision rate coefficient k combines these concepts to describe the frequency of collisions in a gas or plasma, considering the distribution of particle velocities. It is defined as[1]

$$k = \sqrt{\frac{2e}{m}} \int_0^{\infty} E_k \sigma f \mathrm{d}E_k \tag{1.2}$$

where e is the unit charge, m is the particle mass (use reduced mass for multiple species in the collisions), and f is the energy distribution function which describes the probability of particles' energy. The mean energy can be computed as integrating over all possible E_k, to account for the likelihood of collisions at various energies. This statistical approach is crucial for understanding and predicting reaction rates in plasmas, where temperature and kinetic energy distributions significantly influence the collision dynamics. Figure 1.1(B) thus shows an example of the reaction rate coefficients as functions of the mean energy in an oxygen plasma. The attachment reaction is the prementioned $e + O_2 \Rightarrow O^- + O$, and the excitation reactions are denoted with the required collision kinetic energy. The 12.1 eV ionization is $e + O_2 \Rightarrow O_2^+ + 2e$, while the 19.5 eV ionization is the dissociative one: $e + O_2 \Rightarrow O + O^+ + 2e$. One can see that the elastic collision can occur for the particles with all the energies, while all other inelastic collisions have an energy threshold to occur. The ionizations usually need higher energy to occur, while the attachment reaction and the excitations may have non-monotonic relation of their rate coefficients versus the energy. For example, it is obvious that a higher-energy electron is harder for oxygen to capture, which explains the decrement of the attachment rate coefficient at high energy regions.

Overall, plasma chemistry encompasses a range of chemical processes that occur in ionized gases, or plasmas, involving a variety of species including electrons, ions, radicals, and neutral atoms and molecules. The interactions between these species are governed by fundamental physical processes such as collisions. These collisions can be categorized into different types based on their outcomes, which are crucial for understanding plasma behavior and applications in fields like astrophysics, fusion research, and industrial processes. Plasma chemistry in engineering applications is extraordinarily complex due to the multitude of chemical pathways and reactions that can occur simultaneously.[6] In a plasma environment, atoms, molecules, ions, electrons, and radicals interact in various processes, including ionization, excitation, dissociation, recombination, and charge exchange. Each of these processes can lead to different chemical products, creating an intricate network of reaction pathways. For instance, in plasma-assisted combustion, fuel molecules can undergo ionization, leading to the formation of positive ions and electrons.[7,8] These ions can further participate in reactions, such as dissociative recombination, leading to the formation of neutral radicals that can initiate chain reactions in the combustion process. Similarly, in plasma-enhanced chemical vapor deposition (PECVD), precursor gases are broken down into reactive species that form thin films on substrates, with the precise control of film composition and structure requiring careful management of plasma conditions.[9] The complexity of these pathways poses significant challenges in control, optimization, and diagnostics within plasma-based systems. Due to the non-equilibrium nature of plasmas, the populations of various species (electrons, ions, and neutrals) are not governed solely by temperature, but also by the energy distribution functions of the particles. This makes it difficult to predict and control the outcome of the chemical processes.[10] Additionally, the presence of numerous intermediate species and transient states can complicate the optimization of plasma processes for desired outcomes, such as achieving specific film qualities in PECVD or maximizing efficiency in plasma-assisted combustion.[9,11] Diagnosing the plasma state and understanding the underlying chemistry requires advanced diagnostic tools, such as optical emission spectroscopy, mass spectrometry, and laser-induced fluorescence, to monitor species concentrations and energy distributions.[12-14] However, interpreting these diagnostic signals can be challenging due to the overlapping spectral lines and the complex interactions among species. Consequently, developing accurate models and simulations to predict plasma behavior and optimize process parameters remains a critical and challenging task in plasma chemistry for engineering applications.[2,15] In the following sections of this chapter, more detailed information of these applications will be introduced where we will see how plasma chemistry plays a role.

1.2 Space Electric Propulsion

Plasma-based space electric propulsion has emerged as a cutting-edge technology with the potential to revolutionize space exploration.[16–18] By harnessing the unique properties of ionized gases, plasma propulsion systems produced significantly increased efficiency and speed compared to traditional chemical rockets with greater specific impulse.[16–18] Space exploration is relentless in its pursuit of innovation, pushing the boundaries of technology to reach the farthest corners of the cosmos. Electric propulsion systems, such as Hall effect thrusters and ion thrusters (see Fig. 1.2), utilize ionizing neutral gas to generate thrust by accelerating ions to high velocities.[19,20] Unlike conventional chemical propulsion, which relies on combustion and chemical reactions to produce thrust, electric propulsion leverages electromagnetic fields to accelerate ions extracted from a plasma, achieving higher specific impulses I_{sp}. This efficiency makes electric propulsion particularly attractive for deep-space missions, where conserving fuel mass is crucial.[19,20] In these systems, plasma chemistry is fundamental to the generation, maintenance, and manipulation of plasma.[20,21]

Xenon has historically been the most common propellant for electric propulsion subsystems, but has begun to be replaced with more affordable and more readily available alternatives such as krypton.[22,23] A storable alternative to xenon offers advantages on subsystem level by eliminating the need for high-pressure propellant storage.[22] In recent years the use of storable atomic propellants like bismuth and mercury for high thrust applications, and zinc and magnesium for high specific impulse applications has been investigated.[24] Some molecular propellants have been investigated, such as water (where the thruster operates on hydrogen and oxygen gas created by electrolysis),[25] iodine[26] and many others but were all found to have disadvantages such as significantly lower performance compared to xenon. For instance, propellant utilization efficiency of the Hall thruster running with water vapor and oxygen has been assessed to be significantly lower than that of Hall thruster utilizing xenon.[27,28]

There have been various investigations into the use of other molecular gases within Hall thrusters, in particular, within the scope of air breathing electric propulsion.[29,30] They have wide availability and low price and can be regarded as non-toxic. The propellants can be used independently or in a mixture replicating the planetary atmosphere conditions at a certain altitude or within a different planetary atmosphere.[27] In water electrolysis Hall thruster, it is proposed that an electrolyser onboard of the spacecraft could separate water into oxygen (propellant used for the thruster) and hydrogen (for the cathode).[28] Overall, it could be concluded that molecular gases like nitrogen, air, and water vapor are of interest for electric propulsion but, to date, very limited research has been done to understanding ionization efficiency of such propellants.

Taking the Hall thrusters as an example: The thruster operates by creating a radial magnetic field and an axial electric field perpendicular to each other within a discharge chamber[17,24] as shown in Fig. 1.3(A). Propellant gas, typically xenon, is injected into the chamber, where it is ionized by collisions with electrons.[17,24] The electrons are trapped in the magnetic field, circulating as a Hall current in the thruster.[17,20,25] The axial electric field is given by the external cathode and electron plume (a.k.a. neutralizer), it accelerates the positively charged ions towards the exhaust, generating thrust as they exit the thruster at high velocities. The electron plume also neutralizes the ion beam, preventing the buildup of a space charge that would otherwise counteract the thrust. Similarly, as shown in Fig. 1.3(B), an ion thruster pre-ionize the neutral gas by means of electric discharge.[16,31–33] The acceleration stage can be a set of DC biased grid to accelerate the ions while the electrons can be a neutralizer outside of the nozzle. An optional B field can be applied to play a role as a magnetic mirror that can also help to increase the thruster by converting the angular momentum of the ions into their axial momentum.[34]

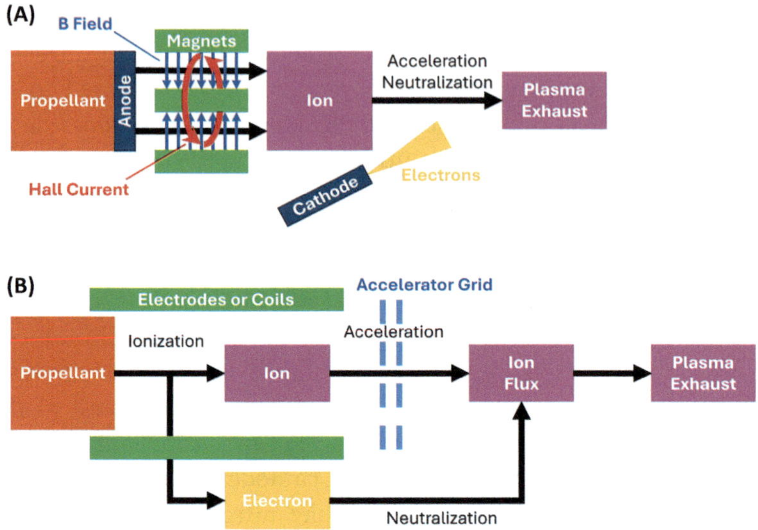

Figure 1.3 The general concept of electric propulsion with two examples: (A) A schematic of Hall thruster and (B) A schematic of ion thruster.

The I_{sp} is a critical parameter in evaluating the performance of electric propulsion systems. It measures the efficiency with which a propulsion system uses propellant, defined as the thrust produced per unit of propellant flow rate, divided by the gravitational constant g. Mathematically, the specific impulse is given by:

$$I_{sp} = \frac{F}{\dot{m}g} \tag{1.3}$$

where F is the thrust and $\dot{m}$ is the mass flow rate of the propellant, as the thrust is directly proportional to the ion mass flow rate and the ion velocity v_i:

$$F = \dot{m}v_i = An_i v_i^2 \tag{1.4}$$

One can see that the ionization rate is a crucial factor influencing the number of ions available for acceleration: A higher ionization rate leads to a greater number of ions, increasing the thrust produced by the thruster in the acceleration stage as shown in Fig. 1.3. Based on the general form of chemical reaction rate equation:

$$\frac{\partial n_\zeta}{\partial t} = \sum_\xi \Xi_{\zeta,\xi} k_\xi \prod_{\xi,\iota} n_{\xi,\iota}^{\Delta_{\xi,\iota}} \quad (1.5)$$

where t is time, n_ζ is density of the ζth species, $\Xi_{\zeta,\xi}$ is coefficient of the ζth species when it is the product in the ξth reaction, k_ξ is reaction rate coefficient of the ξth reaction that can affect the ζth species, and $n_{\xi,\iota}^{\Delta_{\xi,\iota}}$ is density of the ιth reactant in the ξth reaction with a reaction degree of $\Delta_{\xi,\iota}$. Therefore, as an example, the ionization rate of an electron impact ionization e + Xe $\Rightarrow$ Xe^+ + 2e with a rate coefficient k is determined as

$$\frac{\partial n_e}{\partial t} = \frac{\partial n_i}{\partial t} = k n_e n_{Xe} \quad (1.6)$$

where n_e is the electron density and $n_e = n_i$ due to the quasi-neutrality, n_{Xe} is the Xenon density.

However, there are also other reactions such as the recombination: e + Xe^+ $\Rightarrow$ Xe. Therefore, the ion and electron densities cannot be increased with no limit. The ionization and recombination will surely balance each other when the electron density and ion density converged to a certain level. Furthermore, plasma chemistry dictates the production of secondary species, such as metastable states and excited atoms, which can influence the discharge characteristics and stability of the plasma. These species can participate in additional ionization or quenching processes, affecting the electron energy distribution function (EEDF), which is f for the electrons in Eq. (1.2), and, consequently, the ionization efficiency and plasma density.[35,36] Therefore, optimizing electric propulsion systems thus requires a detailed understanding of plasma chemistry, as well as the complex interplay of electromagnetic fields and particle dynamics. Advanced diagnostics and modeling techniques are employed to study the plasma properties, such as ion and electron densities, temperatures, and potential distributions.[37-39] By controlling (including the real-time diagnostics) the plasma conditions, engineers can optimize the efficiency and stability of the propulsion system, enhancing performance and longevity.

However, a better example of thruster with complicated plasma chemistry is the air-breathing electric propulsion, a variation of the space electric propulsion. The air-breathing thruster works near the Kármán line rather than in the deep space and its thrust can cancel out the drag to ensure a long-term fly at such a very low orbit.[29,30] With an inlet at front, the thruster ionizes the air input to generate ion flow, and a high $\dot{m}$ plasma plume at the nozzle. The propellants proposed are oxygen (O_2), carbon dioxide (CO_2) and nitrogen (N_2) as they can be found in various percentages within planetary atmospheres in the solar system.[29,30] Compared with the conventional deep space electric propulsion which has relatively simpler plasma chemistry (elastic collisions, excitations, quenching, ionization, and recombination), N_2, O_2, and other species such as H_2O and CO_2 in the atmosphere can generate hundreds of species through thousands of chemical reactions occurring at the same time.[29,30]

1.3 Plasma-Assisted Ignition and Combustion

Combustion chambers are critical components in a variety of applications, including internal combustion engines, gas turbines, and rocket engines.[40–42] They serve as the environment where fuel and oxidizer mix and react to produce heat, which is then converted into mechanical energy or thrust.[40–42] The fundamental physics of combustion involves the rapid oxidation of a fuel, releasing energy in the form of heat and light. The chemistry of combustion includes a series of exothermic reactions where hydrocarbon fuels react with an oxidizer, typically oxygen, to form combustion products such as water, carbon dioxide, and nitrogen oxides.[43,44] These processes are governed by the conservation of mass, momentum, and energy, as well as the chemical kinetics of the reacting species. Conventional combustion systems face several challenges, including environmental pollution and combustion instability. Environmental concerns are primarily due to the emission of pollutants like carbon monoxide (CO), nitrogen oxides (NOx), unburned hydrocarbons, and particulate matter. These emissions are byproducts of incomplete combustion and high-temperature reactions, particularly in fuel-rich or lean conditions.[45,46] On the other hand, combustion instability can occur due to the coupling of acoustic waves (pressure oscillations), fluid dynamics, and thermal-chemical dynamics.[47,48] This phenomenon, known as thermoacoustic instability, can lead to fluctuations in the combustion process, causing uneven heat release, vibrations, and potential damage to the combustion chamber.[47,48] These instabilities are characterized by pressure and heat release fluctuations that reinforce each other, often described by the Rayleigh criterion, which states that the net work done by pressure fluctuations in phase with heat release leads to instability:[47,48]

$$\int_V \int_0^\tau \frac{1}{\rho c^2} p' \dot{q}' \mathrm{d}t \mathrm{d}V > 0 \tag{1.7}$$

where t and V are the integration unit of time and volume, τ is a period of the thermoacoustic oscillation, ρ is the mass density of the fluid medium, c is the local sound speed, p' is the turbulent component of the pressure wave, and $\dot{q}'$ is the turbulent component of the heat release rate due to the combustion. In this equation, c, ρ, p', and $\dot{q}'$ are all functions of space and time. Plasma-assisted ignition and combustion leverage plasma discharges to enhance and stabilize the combustion process by means of: (1) altering the chemical species and pathways, and (2) altering the fluid and thermal dynamics properties of the gas mixture.[49,50]

It is worth noting that flame itself may be considered as a plasma because a redox reaction may lead to significant free electrons, anions, and cations in gas, and resulting fluid agrees with the definition of plasma prementioned at the beginning of this chapter. Therefore, adding a plasma generator in a combustion chamber, strictly speaking, may be not "generating plasma", but merely increasing the ionization degree significantly. However, the rate coefficients of the reactions in the chemical reaction network are different between the regular combustion and the plasma-involved one as illustrated in Fig. 1.4. The two major redox reactions are

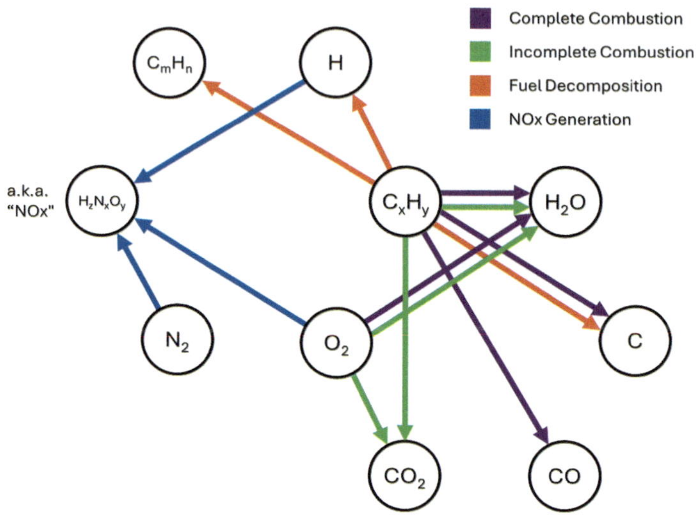

Figure 1.4 Some major reactions in the combustion process, illustrated as a chemical pathway network.

$$C_xH_y + \left(x + \frac{y}{4}\right)O_2 \Rightarrow xCO_2 + \frac{y}{2}H_2O \tag{1.8a}$$

and

$$C_xH_y + \left(x + \frac{y}{4} - z\right)O_2 \Rightarrow xCO + \left(\frac{y}{2} - z\right)H_2O + zC \tag{1.8b}$$

known as the complete and incomplete combustion, while there are also side effects:

$$C_xH_y \Rightarrow C_mH_n + (x-m)C + (y-n)H \tag{1.8c}$$

$$zH + \frac{y}{2}O_2 + \frac{x}{2}N_2 \Rightarrow H_zN_xO_y \tag{1.8d}$$

which are known as the generation of NOx. In Eq. (1.8a–d), the variable x, y, z, m, and n are the numbers of atoms in molecules.

The rate coefficients of them are determined by the Arrhenius equation that is

$$k(T) = A\exp\left(-\frac{E_a}{RT}\right) \tag{1.8}$$

where E_a is the activation energy required to the redox reactions in the combustion chamber, A is a constant amplitude, R is the ideal gas constant, and T is the local reactant temperature. For example, the E_a value for CH_4 is about 175 kJ/mol, and the one for C_8H_{18} is about 125 kJ/mol, and the A values are 1.3×10^{11} cm^3/(mol·s) and 1.0×10^{13} cm^3/(mol·s). In the presence of plasma, additional pathways involving radicals like O, H, and OH can accelerate the combustion process as $H_2 + O \Rightarrow OH + H$ and then $CH_4 + OH \Rightarrow CH_3 + H_2O$. These reactions can significantly alter the overall combustion kinetics, leading to more efficient and complete combustion. The result is a reduction in pollutant formation, such as CO and unburned hydrocarbons, and improved stability of the combustion process. Plasmas generate a high density of radicals (e.g., O, H, OH) and excited species, which lower the activation energy required for combustion reactions. The presence of radicals accelerates the breakdown of fuel molecules, thus promoting faster and more complete combustion. This is a result of merging the CPN as shown in Fig. 1.2 into the CPN in Fig. 1.4.

In addition to plasma chemistry, the increment of ionization degree also means increasing the probability of Coulomb collisions which means charged particles can collide through long-range Coulomb force. This increases the viscosity, thermal conduction rate, and the damping of the sound waves (pressure waves). This is a modification to the macroscopic physical properties of the entire fluid from its microscopic basis.

The above are the physical and chemical mechanisms of how a plasma can be used to control and attenuate combustion instabilities by modulating the reaction rates and the distribution of reactive species. Plasma-assisted ignition and combustion offer a powerful means to enhance the efficiency, stability, and environmental performance of combustion systems. By leveraging the unique properties of plasmas to modify chemical reaction pathways and control instabilities, this technology provides promising solutions for advanced propulsion and energy systems.

1.4 Material Processing

There are many very different types of material processing which heavily rely on plasma chemistry, each playing a crucial role in various industries, particularly in semiconductor manufacturing, surface treatment, and thin-film deposition.[51-53] For example, plasma etching is a critical process in semiconductor fabrication, used to selectively remove material from the surface of a substrate, typically silicon, to create intricate patterns required for integrated circuits.[51,54–57] In this process, a plasma is generated using gases such as chlorine (Cl_2), fluorine (F_2), or oxygen (O_2), which produces reactive ions and radicals.[51] These reactive species chemically interact with the material on the substrate, forming volatile compounds that are then removed, leaving behind the desired pattern, and the precision and control offered by plasma etching make it indispensable for defining features at the nanoscale.[51] Another example is Plasma-Enhanced Atomic Layer Deposition (PEALD), which is a variant of Atomic Layer Deposition (ALD) that incorporates plasma to enhance the surface reactions during film growth, while atomic layer etching is the opposite.[58,59] ALD is a technique for depositing extremely thin and uniform films, one atomic layer at a time.[58] In PEALD, the plasma serves to activate the surface and drive the reaction between the precursor and the surface, allowing for deposition at lower temperatures compared to thermal ALD.[60–62] This is particularly useful for depositing materials like oxides, nitrides, and metals with high conformality and precise thickness control, even on complex 3D structures. Ceramics and other compounds are also usually deposited in industry via magnetron sputtering and vacuum cathodic arc discharge using reactive atmospheres. These physical vapor deposition (PVD) techniques allow for an accurate control of film thickness and physical-chemical properties, such as stoichiometry and microstructure, to prepare surfaces for specific mechanical, optical, and electrical applications.[63] Similarly, Plasma Immersion Ion Implantation (PIII) is a technique used to modify the surface properties of materials by implanting ions into the substrate using plasma.[64] Unlike traditional ion implantation, which requires an energetic ion beam, PIII immerses the entire substrate in a plasma, and a high-voltage pulse is applied to the substrate to accelerate ions from the plasma into the material.[64] This method is widely used to improve surface hardness, corrosion resistance, and biocompatibility of materials, and it is particularly effective for treating complex-shaped objects. Plasma chemistry is also used for plasma surface functionalization, a technique used to modify the surface chemistry of materials to enhance properties such as adhesion, wettability, and biocompatibility.[65–68] In this process, a plasma, often generated from gases like oxygen (O_2), nitrogen (N_2), or ammonia (NH_3), is used to create reactive species that interact with the surface. This can lead to the introduction of functional groups, such as hydroxyl (-OH) or amine (-NH_2) groups, onto the surface, making it more chemically active. This technique is commonly used in the preparation of surfaces for bonding, painting, or coating.[65–68]

While there are many other examples such as plasma cleaning and plasma nitriding and carburizing, we will focus on one of the most important examples of plasma-based material processing to show how plasma chemistry plays a role: Plasma-Enhanced Chemical Vapor Deposition (PECVD) as a critical subroutine in the manufacturing of semiconductors.[69–73] The semiconductor industry is at the heart of modern electronics, powering everything from smartphones to supercomputers. The manufacturing process of semiconductors is a highly intricate series of steps that involve the creation of extremely thin layers of materials on a substrate, usually silicon, to build up complex structures such as transistors, capacitors, and interconnects. These structures are defined at the nanoscale, requiring precise control over material deposition, etching, doping, and patterning. Among these processes, chemical vapor deposition (CVD) plays a critical role in forming high-quality thin films with desired electrical, optical, and mechanical properties. One specialized form of CVD, known as PECVD, is widely used in semiconductor manufacturing due to its ability to deposit thin films at relatively low temperatures while maintaining excellent film quality.[69–73] A typical PECVD reaction chamber and the major CPN of it is shown in Fig. 1.5. An example is the deposition of SiO_2 on the Si substrate. Please note that there are many other PECVD examples such as the deposition of fluorine-containing films.

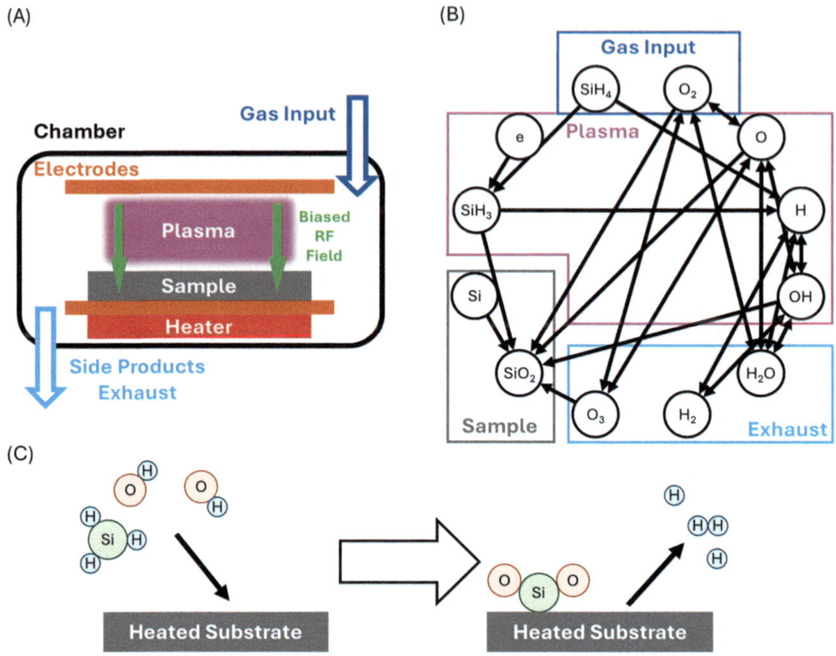

Figure 1.5 The main idea of PECVD to deposit SiO_2 as an example where (A) is the schematic of a PECVD reactor chamber and (B) is a major part of its CPN. Please note that the ions, excited species, photons, and electron impacts are not shown in the CPN. Nitrogen elements may also be added in some cases.

In the example shown in Fig. 1.5, the PECVD process relies on the generation of plasma within a reaction chamber, where the substrate is placed. Plasma is created by applying a radiofrequency (RF) electric field, typically in the range of 13.56 MHz, across a low-pressure gas mixture containing the precursor gases. This RF field ionizes the gas, creating a plasma composed of electrons, ions, radicals, and neutral atoms. The energy distribution within the plasma is non-thermal, meaning that the electrons can reach temperatures of 1–10 eV, while the bulk gas remains relatively cool at around 300–600 K.[71,73] The high-energy electrons in the plasma are crucial for dissociating the precursor molecules into reactive species. The chemical reactions involved are complex and depend on the specific precursors used, but they generally follow pathways that are significantly enhanced by the presence of plasma. The kinetic energy gained by ions in the plasma sheath, typically ranging from a few to hundreds of eV, plays a critical role in determining the properties of the deposited film. Ions can assist in the densification of the film by providing additional energy to the surface, which promotes better atomic arrangement and reduces voids. However, excessive ion energy can lead to damage or unwanted etching of the substrate, so controlling the ion energy is essential for optimizing the PECVD process.

PECVD is a versatile and essential technique in semiconductor manufacturing, particularly for depositing insulating and passivating layers, such as silicon dioxide (SiO_2), silicon nitride (Si_3N_4), and amorphous silicon.[74–76] An example of SiO_2 deposition is shown in Fig. 1.5(C). In a standard CVD process, a substrate is exposed to volatile precursors that react or decompose on the surface, leading to film growth. However, traditional CVD processes often require high temperatures to drive these reactions, which can be detrimental to other materials or structures already present on the wafer.[74–76] PECVD addresses this limitation by introducing plasma, an ionized gas, to activate the chemical reactions at much lower temperatures, typically below 400°C. This makes PECVD especially valuable for fabricating complex semiconductor devices where thermal budgets are strictly controlled.[74–76] In PECVD, plasma provides the necessary energy to dissociate the precursor gases into reactive radicals, ions, and other species, which then react on the substrate to form a thin film. The ability to deposit films at low temperatures while still achieving good film uniformity, density, and step coverage makes PECVD indispensable for processes such as the deposition of dielectric layers in integrated circuits, passivation layers in photovoltaic cells, and barrier layers in advanced packaging technologies.

1.5 Biomedicine and Agriculture

Plasma-based biomedicine is an emerging interdisciplinary field that explores the use of non-thermal atmospheric pressure plasmas for therapeutic and agricultural applications.[77–79] The plasma is also known as low-temperature atmospheric plasma or cold atmospheric plasma (CAP). Unlike the high-temperature plasmas used in industrial processes, the CAP operates at temperatures close to room temperature, making them safe for direct interaction with biological tissues. The potential of plasma medicine spans a wide range of applications, including wound healing, cancer treatment, sterilization, and even stimulation of plant growth in agriculture. The efficacy of plasma in these areas is largely due to its ability to generate a rich mixture of reactive species that interact with cells, tissues, and microorganisms, triggering beneficial biological responses.[80,81]

When plasma is generated at atmospheric pressure, it produces a complex mixture of reactive species, including reactive oxygen species (ROS), reactive nitrogen species (RNS), charged particles, UV radiation, and electromagnetic fields.[82–85] These species are produced through various chemical pathways that occur in both the gas phase (air) and liquid phase (medium) environments. The interaction of plasma with air primarily leads to the formation of ROS such as ozone (O_3), hydroxyl radicals (•OH), superoxide anions (O_2^-), and singlet oxygen ($^1O_2$).[86–92] Additionally, RNS such as nitric oxide (NO) and peroxynitrite ($ONOO^-$) are generated through reactions involving nitrogen and oxygen species in the air.[86–92] In vitro, when plasma interacts with biological media (e.g., saline solutions, culture media), these reactive species dissolve into the liquid, leading to further chemical transformations. For example, in aqueous solutions, ROS such as hydrogen peroxide (H_2O_2) and nitrite (NO_2^-) can be formed.[86–92] These species are capable of diffusing through the medium and interacting with cells, leading to various biological effects. The production and concentration of these reactive species can be finely tuned by adjusting plasma parameters such as power, gas composition, and exposure time, allowing for targeted therapeutic effects.[86–92]

The reactive species generated by plasma can penetrate biological barriers and interact with cells both at the surface (intercellular) and within the cytoplasm and organelles (intra-cellular). Extracellularly, ROS and RNS can modify the extracellular matrix, influence cell-cell signaling, and trigger surface receptor-mediated pathways.[93–95] For instance, the interaction of ROS with cell membrane lipids can lead to lipid peroxidation, altering membrane fluidity and affecting the function of membrane-bound receptors and ion channels.[93–95] Intracellularly, reactive species can enter cells through diffusion or active transport mechanisms. Once inside, they can interact with various biomolecules, including proteins, nucleic acids, and lipids.[93–95] ROS can oxidize amino acid residues in proteins, leading to conformational changes or inactivation, while RNS can also modify protein function.[96,97] These modifications can influence cellular signaling pathways, such as those mediated by the mitogen-activated protein kinase (MAPK) and nuclear factor kappa-light-chain-enhancer of activated B cells (NF-κB), which are involved in processes like cell proliferation, apoptosis, and inflammation.[80,84,97] Oxidation of protein polymers that make up the cytoskeleton, such as actin filaments or microtubules, that can also result in changes in cell morphology and physiology. Changes in cytoskeletal organization can be caused by direct oxidation by exogenous RONS, or secondary effects from increased intracellular calcium concentration and altered mitochondrial function.[95,98] One of the most critical intracellular targets of plasma-generated ROS and RNS is DNA.[95,98] While low levels of reactive species can induce beneficial stress responses, such as the activation of antioxidant defense mechanisms and DNA repair pathways, high levels can lead to oxidative damage, resulting in strand breaks, base modifications, cell cycle arrest, and ultimately, cell death. This dual role of plasma in modulating cellular responses makes it a powerful tool for selectively targeting diseased cells, such as cancer cells, while sparing healthy tissue.[94,95,99–101]

Especially, plasma treatment in cancer therapy relies on the generation of a high concentration of reactive species, which can penetrate cancer cells and induce oxidative stress.[102] Once inside the cell, ROS and RNS, such as hydrogen peroxide (H_2O_2), superoxide anions (O_2^-), and nitric oxide (NO), can interact with cellular components, leading to the oxidation of lipids, proteins, and DNA.[85,86,88,103] This oxidative damage disrupts critical cellular functions, leading to apoptosis (programmed cell death) or necrosis.[84,93,104–108] ROS can activate pro-apoptotic pathways, such as the intrinsic (mitochondrial) apoptotic pathway, where oxidative damage to the mitochondria leads to the release of cytochrome c and the activation of caspases, which are enzymes that execute cell death. In addition to the chemical effects, plasma treatment also exerts physical effects on cancer cells.[109–112] The electromagnetic fields, including radiofrequency (RF) emissions, generated by plasma can influence cellular membrane potentials and ion channels.[109–112] This can lead to the disruption of ion homeostasis, particularly calcium ion (Ca^{2+}) signaling, which is critical for maintaining cellular integrity and function. RF emissions may also induce localized heating and mechanical stress, contributing to the destabilization of cancer cells. Therefore, since the plasma used for biomedicine must be a low temperature one exposed to the atmosphere, while both its physical emissions and chemical species are used, the following types of plasma generators had been reported and studied over a decade as shown in Fig. 1.6 of plasma jet and Fig. 1.7 of dielectric barrier discharge (DBD) generator.

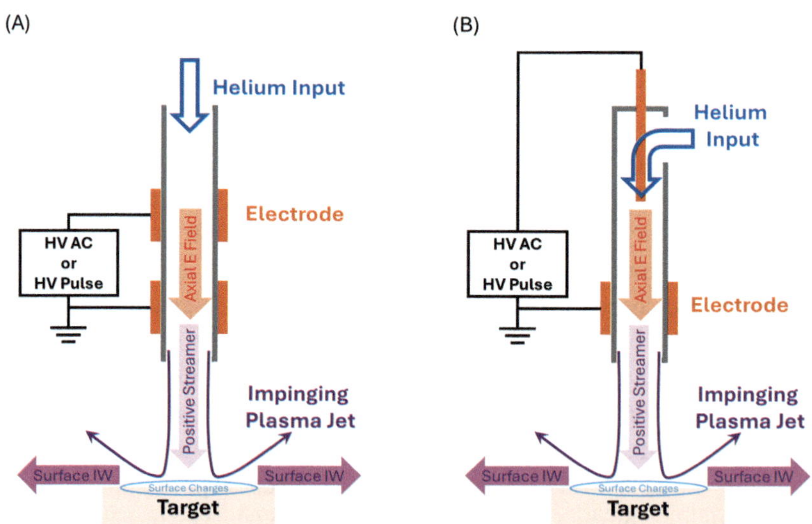

Figure 1.6 The schematics of two most popular designs of plasma jet: (A) the DBD-like plasma jet and (B) the conventional pin-ring plasma jet.

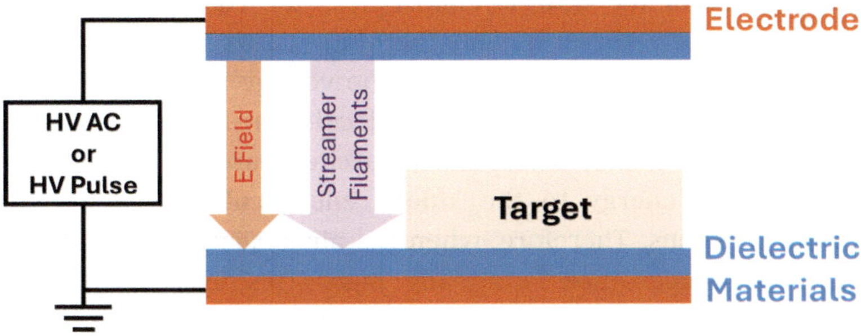

Figure 1.7 The schematic of a typical DBD generator.

As shown in Fig. 1.6(A), two ring electrodes, such as copper tapes, are wrapped around a dielectric tube which can guide a noble gas flow, usually helium to flow through and reach the target exposed in open air. An axial electric field is thus built between the electrodes, leading to a streamer propagation which is also known as an axial ionization wave (IW).[113–118] The ionization wave will then propagate along the helium flow and reach the outside area of the nozzle. Using noble gas to guide the streamer discharge is also the main reason of plasma jet's low temperature. Taking helium as an example, the 1st ionization threshold is 24.58 eV which is higher than the values of nitrogen and oxygen in the air. However, because N_2 and O_2 are diatomic molecules with more degrees of freedom such as the vibrational and rotational excitations, the helium can actually have a higher ionization rate compared with N_2 and O_2 with the same EEDF. Therefore, when the discharge voltage (voltage between the two ring electrodes) is low but enough to ionize helium, it cannot ionize the air. When the helium is ionized, more free electrons are generated during the avalanche, and this can increase the ionization rate of air (see Eq. 1.5 and Eq. 1.6 where increasing the reactant can increase the rate even though the rate coefficient is low) and finally, the air can be significantly ionized. As a result, the electron temperature is high because it is accelerated by the axial electric field and the streamer head electric field and it can ionize other particles during the collisions, while the atoms, molecules, and ions are at low thermal and drift velocity because of their heavy mass. This leads to the low-temperature plasma so that the bulk temperature of the flow is almost near the room temperature. However, when there is only air, one needs a much higher discharge voltage to ionize. Before the voltage reaches the value required, many of the N_2 and O_2 molecules are at the vibrational and rotational excited states. These excited particles can transfer their excited energy to the kinetic energy of thermal motion during collisions. Therefore, when the air is significantly ionized, the bulk temperature is already high. This leads to a high-temperature plasma, such as arc, for the atmospheric air when there is no helium. Of course, for biomedical purpose, we therefore must use noble gas guided plasma. In Fig. 1.6(B), another commonly used design is to replace the upper ring electrode with the pin electrode inside the tube. This means a direct contact between the pin electrode and the plasma, leading to a higher discharge current but it requires a relatively lower discharge voltage because the gas can get closer to the electrode tip where the electric field is much stronger considering the highly non-linear field.

Once the streamer head is approaching the target, its space charge electric field will lead to a surface charge distribution at the surface of the target.[113-118] Especially the liquid target can therefore present a spatial distribution of excited species when the streamer contacts the surface and the brightness pattern is known as self-organization pattern, because the formation of charges and excited species can be considered as a self-organization process.[119-121] If the target has a significant dielectric property, the surface interaction does not consume much of the streamer energy and as a result, the streamer will not end at the contact spot, but propagate further along the surface of the target, also known as surface ionization wave (SIW), where the helium flow is forming an impinging jet.[122-124] The streamer in air thus generates ROS and RNS, along with the RF emissions to treat the target.

DBD is another commonly used form of CAP generated by applying an alternating high voltage across two electrodes, with at least one of the electrodes covered by a dielectric material such as glass, ceramic, or polymer.[125-132] This configuration creates a strong electric field that ionizes the surrounding gas, generating plasma that is rich in reactive species, including ions, electrons, radicals, and excited molecules.[129] The dielectric barrier prevents a continuous arc discharge by not only limiting the current flow, but also using noble gas. DBD contains multiple micro-discharges across the gas gap, each of them is also a small-size streamer.[129] The advantage of using DBD is clearly a larger covering area compared to the plasma jets, while the latter has an advantage that does not require another electrode behind the target.

The CPN of both is mainly adding the electron-He collisions to the CPN in Fig. 1.2. According to some of the latest research works, there are more than 100 different species (including photons with different wavelengths, and excited species at different states), and more than 1000 chemical reactions are counted in the CPN of CAP. To accurately diagnose and control the CAP, machine learning is now the latest and most effective method which is also currently under further study and development. These topics will be discussed in the next chapter.

References

1. Hagelaar, G. J. M.; Pitchford, L. C. Solving the Boltzmann Equation to Obtain Electron Transport Coefficients and Rate Coefficients for Fluid Models. *Plasma Sources Sci. Technol.* **2005**, *14* (4), 722–733. https://doi.org/10.1088/0963-0252/14/4/011.
2. Sakiyama, Y.; Graves, D. B.; Chang, H.-W.; Shimizu, T.; Morfill, G. E. Plasma Chemistry Model of Surface Microdischarge in Humid Air and Dynamics of Reactive Neutral Species. *J. Phys. Appl. Phys.* **2012**, *45* (42), 425201. https://doi.org/10.1088/0022-3727/45/42/425201.
3. Gutcheck, R. A.; Zipf, E. C. Excitation of the CO Fourth Positive System by the Dissociative Recombination of CO_2^+ Ions. *J. Geophys. Res.* **1973**, *78* (25), 5429–5436. https://doi.org/10.1029/JA078i025p05429.
4. Johnson, A. W.; Fowler, R. G. Measured Lifetimes of Rotational and Vibrational Levels of Electronic States of N_2. *J. Chem. Phys.* **1970**, *53* (1), 65–72. https://doi.org/10.1063/1.1673834.
5. Stevens, M. H.; Conway, R. R. Calculated OH $A^2\ \Sigma^+ - X^2\ \Pi$ (0,0) Band Rotational Emission Rate Factors for Solar Resonance Fluorescence. *J. Geophys. Res. Atmospheres* **1999**, *104* (D13), 16369–16378. https://doi.org/10.1029/1998JD100073.

6. Lin, L.; Yan, D.; Lee, T.; Keidar, M. Self-Adaptive Plasma Chemistry and Intelligent Plasma Medicine. *Adv. Intell. Syst.* **2022**, *4* (3), 2100112. https://doi.org/10.1002/aisy.202100112.
7. Bulat, M. P.; Bulat, P. V.; Denissenko, P. V.; Esakov, I. I.; Grachev, L. P.; Lavrov, P. V.; Volkov, K. N.; Volobuev, I. A. Plasma-Assisted Ignition and Combustion of Lean and Rich Air/Fuel Mixtures in Low- and High-Speed Flows. *Acta Astronaut.* **2020**, *176*, 700–709. https://doi.org/10.1016/j.actaastro.2020.04.028.
8. Deak, N.; Bellemans, A.; Bisetti, F. Plasma-Assisted Ignition of Methane/Air and Ethylene/Air Mixtures: Efficiency at Low and High Pressures. *Proc. Combust. Inst.* **2021**, *38* (4), 6551–6558. https://doi.org/10.1016/j.proci.2020.06.126.
9. Meyyappan, M.; Delzeit, L.; Cassell, A.; Hash, D. Carbon Nanotube Growth by PECVD: A Review. *Plasma Sources Sci. Technol.* **2003**, *12* (2), 205–216. https://doi.org/10.1088/0963-0252/12/2/312.
10. Kim, B.; Park, K.; Lee, D. Use of Neural Network to Model the Deposition Rate of PECVD-Silicon Nitride Films. *Plasma Sources Sci. Technol.* **2005**, *14* (1), 83–88. https://doi.org/10.1088/0963-0252/14/1/011.
11. Choe, M. S.; Lee, K. T.; Kim, K. S.; Choi, D. S. Effect of the Plasma Jet Ignition and Flame Kernel Under the Combustion Process in a Constant Volume Combustion Chamber. *Int. J. Automot. Technol.* **2020**, *21* (4), 833–842. https://doi.org/10.1007/s12239-020-0081-0.
12. Riès, D.; Dilecce, G.; Robert, E.; Ambrico, P. F.; Dozias, S.; Pouvesle, J.-M. LIF and Fast Imaging Plasma Jet Characterization Relevant for NTP Biomedical Applications. *J. Phys. Appl. Phys.* **2014**, *47* (27), 275401. https://doi.org/10.1088/0022-3727/47/27/275401.
13. Elliott, S.; Lax, P.; Leonov, S. B.; Carter, C.; Ombrello, T. Acetone PLIF Visualization of the Fuel Distribution at Plasma-Enhanced Supersonic Combustion. *Exp. Therm. Fluid Sci.* **2022**, *136*, 110668. https://doi.org/10.1016/j.expthermflusci.2022.110668.
14. Settersten, T. B.; Dreizler, A.; Farrow, R. L. Temperature- and Species-Dependent Quenching of CO *B* Probed by Two-Photon Laser-Induced Fluorescence Using a Picosecond Laser. *J. Chem. Phys.* **2002**, *117* (7), 3173–3179. https://doi.org/10.1063/1.1493198.
15. Pancheshnyi, S.; Biagi, S.; Bordage, M. C.; Hagelaar, G. J. M.; Morgan, W. L.; Phelps, A. V.; Pitchford, L. C. The LXCat Project: Electron Scattering Cross Sections and Swarm Parameters for Low Temperature Plasma Modeling. *Chem. Phys.* **2012**, *398*, 148–153. https://doi.org/10.1016/j.chemphys.2011.04.020.

16. Burton, R. L.; Turchi, P. J. Pulsed Plasma Thruster. *J. Propuls. Power* **1998**, *14* (5), 716–735. https://doi.org/10.2514/2.5334.

17. Levchenko, I.; Xu, S.; Teel, G.; Mariotti, D.; Walker, M. L. R.; Keidar, M. Recent Progress and Perspectives of Space Electric Propulsion Systems Based on Smart Nanomaterials. *Nat. Commun.* **2018**, *9* (1), 879. https://doi.org/10.1038/s41467-017-02269-7.

18. Zolotukhin, D. B.; Daniels, K. P.; Bandaru, S. R. P.; Keidar, M. Magnetoplasmadynamic Two-Stage Micro-Cathode Arc Thruster for CubeSats. *Plasma Sources Sci. Technol.* **2019**, *28* (10), 105001. https://doi.org/10.1088/1361-6595/ab4170.

19. Jorns, B. A.; Byrne, M. P. Model for the Dependence of Cathode Voltage in a Hall Thruster on Facility Pressure. *Plasma Sources Sci. Technol.* **2021**, *30* (1), 015012. https://doi.org/10.1088/1361-6595/abd3b6.

20. Keidar, M.; Boyd, I. D.; Beilis, I. I. Plasma Flow and Plasma–Wall Transition in Hall Thruster Channel. *Phys. Plasmas* **2001**, *8* (12), 5315–5322. https://doi.org/10.1063/1.1421370.

21. Meyer, M. E.; Byrne, M. P.; Boyd, I. D.; Jorns, B. A. Quantifying Uncertainty in Predictions of Spacecraft Erosion Induced by a Hall Thruster. *J. Spacecr. Rockets* **2022**, *59* (3), 988–1000. https://doi.org/10.2514/1.A35196.

22. Linnell, J. A.; Gallimore, A. D. Efficiency Analysis of a Hall Thruster Operating with Krypton and Xenon. *J. Propuls. Power* **2006**, *22* (6), 1402–1418. https://doi.org/10.2514/1.19613.

23. Bugrova, A. I.; Bishaev, A. M.; Desyatskov, A. V.; Kozintseva, M. V.; Lipatov, A. S.; Dudeck, M. Experimental Investigations of a Krypton Stationary Plasma Thruster. *Int. J. Aerosp. Eng.* **2013**, *2013*, 1–7. https://doi.org/10.1155/2013/686132.

24. Kieckhafer, A.; King, L. B. Energetics of Propellant Options for High-Power Hall Thrusters. *J. Propuls. Power* **2007**, *23* (1), 21–26. https://doi.org/10.2514/1.16376.

25. Schwertheim, A.; Knoll, A. Low Power Thrust Measurements of the Water Electrolysis Hall Effect Thruster. *CEAS Space J.* **2022**, *14* (1), 3–17. https://doi.org/10.1007/s12567-021-00350-y.

26. Szabo, J.; Pote, B.; Paintal, S.; Robin, M.; Hillier, A.; Branam, R. D.; Huffmann, R. E. Performance Evaluation of an Iodine-Vapor Hall Thruster. *J. Propuls. Power* **2012**, *28* (4), 848–857. https://doi.org/10.2514/1.B34291.

27. Tirila, V.-G.; Demairé, A.; Ryan, C. N. Review of Alternative Propellants in Hall Thrusters. *Acta Astronaut.* **2023**, *212*, 284–306. https://doi.org/10.1016/j.actaastro.2023.07.047.

28. Tejeda, J. M.; Knoll, A. A Water Vapour Fuelled Hall Effect Thruster: Characterization and Comparison with Oxygen. *Acta Astronaut.* **2023**, *211*, 702–715. https://doi.org/10.1016/j.actaastro.2023.06.046.

29. Taploo, A.; Lin, L.; Keidar, M. Air Ionization in Self-Neutralizing Air-Breathing Plasma Thruster. *J. Electr. Propuls.* **2022**, *1* (1), 25. https://doi.org/10.1007/s44205-022-00022-x.

30. Taploo, A.; Lin, L.; Keidar, M. Analysis of Ionization in Air-Breathing Plasma Thruster. *Phys. Plasmas* **2021**, *28* (9), 093505. https://doi.org/10.1063/5.0059896.

31. Zhuang, T.; Shashurin, A.; Denz, T.; Keidar, M.; Vail, P.; Pancotti, A. Performance Characteristics of Micro-Cathode Arc Thruster. *J. Propuls. Power* **2014**, *30* (1), 29–34. https://doi.org/10.2514/1.B34567.

32. Zolotukhin, D. B.; Hurley, S.; Keidar, M. Anode Ablation and Performance Improvement of Microcathode Arc Thruster. *Plasma Sources Sci. Technol.* **2019**, *28* (3), 034001. https://doi.org/10.1088/1361-6595/ab01ec.

33. Zolotukhin, D. B.; Tyunkov, A. V.; Yushkov, Y. G.; Oks, E. M.; Keidar, M. Improvement of Microcathode Arc Thruster Lifetime by Deposition of Boron-Containing Coating. *J. Propuls. Power* **2020**, *36* (5), 744–751. https://doi.org/10.2514/1.B37790.

34. Keidar, M.; Schein, J.; Wilson, K.; Gerhan, A.; Au, M.; Tang, B.; Idzkowski, L.; Krishnan, M.; Beilis, I. I. Magnetically Enhanced Vacuum Arc Thruster. *Plasma Sources Sci. Technol.* **2005**, *14* (4), 661–669. https://doi.org/10.1088/0963-0252/14/4/004.

35. Banks, P. Collision Frequencies and Energy Transfer Electrons. *Planet. Space Sci.* **1966**, *14* (11), 1085–1103. https://doi.org/10.1016/0032-0633(66)90024-9.

36. Margenau, H. Theory of High Frequency Gas Discharges. I. Methods for Calculating Electron Distribution Functions. *Phys. Rev.* **1948**, *73* (4), 297–308. https://doi.org/10.1103/PhysRev.73.297.

37. Muraoka, K.; Kono, A. Laser Thomson Scattering for Low-Temperature Plasmas. *J. Phys. Appl. Phys.* **2011**, *44* (4), 043001. https://doi.org/10.1088/0022-3727/44/4/043001.

38. Dzierżęga, K.; Mendys, A.; Pokrzywka, B.; Zawadzki, W.; Pellerin, S. Simultaneous Measurement of Electron and Heavy Particle

Temperatures in He Laser-Induced Plasma by Thomson and Rayleigh Scattering. *Appl. Phys. Lett.* **2013**, *102* (13), 134108. https://doi.org/10.1063/1.4801467.

39. Shneider, M. N.; Miles, R. B. Microwave Diagnostics of Small Plasma Objects. *J. Appl. Phys.* **2005**, *98* (3), 033301. https://doi.org/10.1063/1.1996835.
40. Petrescu, R. V.; Aversa, R.; Akash, B.; Bucinell, R.; Corchado, J.; Berto, F.; Mirsayar, M.; Apicella, A.; Petrescu, F. I. T. Modern Propulsions for Aerospace-A Review. *J. Aircr. Spacecr. Technol.* **2017**, *1* (1), 1–8. https://doi.org/10.3844/jastsp.2017.1.8.
41. Casiano, M. J.; Hulka, J. R.; Yang, V. Liquid-Propellant Rocket Engine Throttling: A Comprehensive Review. *J. Propuls. Power* **2010**, *26* (5), 897–923. https://doi.org/10.2514/1.49791.
42. Miller, W. H.; Barrington, D. K. A Review of Contemporary Solid Rocket Motor Performance Prediction Techniques. *J. Spacecr. Rockets* **1970**, *7* (3), 225–237. https://doi.org/10.2514/3.29911.
43. Serbin, S. I.; Vilkul, S. V.; Patlaichuk, V. M. Improving the Efficiency of Fuel Combustion in a 2.5-MW Contact-Type Gas Turbine Engine Using the Cross-Flow Plasma-Fuel Nozzle. *IEEE Trans. Plasma Sci.* **2022**, *50* (6), 1675–1680. https://doi.org/10.1109/TPS.2022.3173561.
44. Kroniger, D.; Horikawa, A.; Funke, H. H.-W.; Pfaeffle, F.; Kishimoto, T.; Okada, K. Experimental and Numerical Investigation on the Effect of Pressure on Micromix Hydrogen Combustion. In *Volume 3A: Combustion, Fuels, and Emissions*; American Society of Mechanical Engineers: Virtual, Online, 2021; p. V03AT04A025. https://doi.org/10.1115/GT2021-58926.
45. Kim, K.; Choi, D. Research on the Reaction Progress of Thermodynamic Combustion Based on Arc and Jet Plasma Energies Using Experimental and Analytical Methods. *J. Mech. Sci. Technol.* **2018**, *32* (4), 1869–1878. https://doi.org/10.1007/s12206-018-0343-1.
46. Li, X.; He, B.-Q.; Zhao, H. Kinetics of Low-Temperature Plasma-Assisted Ignition of Ethanol-Gasoline Surrogate under Gasoline Engine like Conditions. *Combust. Sci. Technol.* **2023**, *195* (12), 2750–2773. https://doi.org/10.1080/00102202.2022.2040998.
47. Beita, J.; Talibi, M.; Sadasivuni, S.; Balachandran, R. Thermoacoustic Instability Considerations for High Hydrogen Combustion in Lean Premixed Gas Turbine Combustors: A Review. *Hydrogen* **2021**, *2* (1), 33–57. https://doi.org/10.3390/hydrogen2010003.

48. Singh, P.; Mishra, A.; Pal, S.; Malhi, G. S. Overview of Thermoacoustic Instability in Gas Turbine Combustion Chamber. **2020**, *29* (10).

49. Deng, K.; Zhao, S.; Xue, C.; Hu, J.; Zhong, Y.; Zhong, Y. Combustion Instability of Swirl Premixed Flame with Dielectric Barrier Discharge Plasma. *Processes* **2021**, *9* (8), 1405. https://doi.org/10.3390/pr9081405.

50. Zhou, S.; Nie, W.; Tian, Y. High Frequency Combustion Instability Control by Discharge Plasma in a Model Rocket Engine Combustor. *Acta Astronaut.* **2021**, *179*, 391–406. https://doi.org/10.1016/j.actaastro.2020.11.010.

51. Donnelly, V. M.; Kornblit, A. Plasma Etching: Yesterday, Today, and Tomorrow. *J. Vac. Sci. Technol. Vac. Surf. Films* **2013**, *31* (5), 050825. https://doi.org/10.1116/1.4819316.

52. Nie, Q.; Cao, Z.; Ren, C. S.; Wang, D. Z.; Kong, M. G. A Two-Dimensional Cold Atmospheric Plasma Jet Array for Uniform Treatment of Large-Area Surfaces for Plasma Medicine. *New J. Phys.* **2009**, *11* (11), 115015. https://doi.org/10.1088/1367-2630/11/11/115015.

53. Lien, S.-Y.; Chang, Y.-Y.; Cho, Y.-S.; Wang, J.-H.; Weng, K.-W.; Chao, C.-H.; Chen, C.-F. Characterization of HF-PECVD a-Si:H Thin Film Solar Cells by Using OES Studies. *J. Non-Cryst. Solids* **2011**, *357* (1), 161–164. https://doi.org/10.1016/j.jnoncrysol.2010.09.044.

54. Bouchilaoun, M.; Soltani, A.; Chakroun, A.; Jaouad, A.; Darnon, M.; Boone, F.; Maher, H. A Hydrogen Plasma Treatment for Soft and Selective Silicon Nitride Etching. *Phys. Status Solidi A* **2018**, *215* (9), 1700658. https://doi.org/10.1002/pssa.201700658.

55. Jeong, J. Y.; Babayan, S. E.; Tu, V. J.; Park, J.; Henins, I.; Hicks, R. F.; Selwyn, G. S. Etching Materials with an Atmospheric-Pressure Plasma Jet. *Plasma Sources Sci. Technol.* **1998**, *7* (3), 282–285. https://doi.org/10.1088/0963-0252/7/3/005.

56. Racka-Szmidt, K.; Stonio, B.; Żelazko, J.; Filipiak, M.; Sochacki, M. A Review: Inductively Coupled Plasma Reactive Ion Etching of Silicon Carbide. *Materials* **2021**, *15* (1), 123. https://doi.org/10.3390/ma15010123.

57. Veilleux, J. M.; El-Genk, M. S.; Chamberlin, E. P.; Munson, C.; FitzPatrick, J. Etching of UO_2 in NF_3 RF Plasma Glow Discharge. *J. Nucl. Mater.* **2000**.

58. Rahman, F.; Runyon, J. C. Atomic Layer Processes for Material Growth and Etching—A Review. *IEEE Trans. Semicond. Manuf.* **2021**, *34* (4), 500–512. https://doi.org/10.1109/TSM.2021.3112502.

59. Oehrlein, G. S.; Metzler, D.; Li, C. Atomic Layer Etching at the Tipping Point: An Overview. *ECS J. Solid State Sci. Technol.* **2015**, *4* (6), N5041–N5053. https://doi.org/10.1149/2.0061506jss.

60. Ji, Y. J.; Kim, K. S.; Kim, K. H.; Byun, J. Y.; Yeom, G. Y. A Brief Review of Plasma Enhanced Atomic Layer Deposition of Si_3N_4. *Appl. Sci. Converg. Technol.* **2019**, *28* (5), 142–147. https://doi.org/10.5757/ASCT.2019.28.5.142.

61. Knoops, H. C. M.; Faraz, T.; Arts, K.; Kessels, W. M. M. (Erwin). Status and Prospects of Plasma-Assisted Atomic Layer Deposition. *J. Vac. Sci. Technol. Vac. Surf. Films* **2019**, *37* (3), 030902. https://doi.org/10.1116/1.5088582.

62. Kim, H.; Oh, I.-K. Review of Plasma-Enhanced Atomic Layer Deposition: Technical Enabler of Nanoscale Device Fabrication. *Jpn. J. Appl. Phys.* **2014**, *53* (3S2), 03DA01. https://doi.org/10.7567/JJAP.53.03DA01.

63. Ohring, M. Thin-Film Evaporation Processes. *Mater. Sci. Thin Films* **2002**, 95–144.

64. Anders, A.; Anders, S.; Brown, I. G.; Dickinson, M. R.; MacGill, R. A. Metal Plasma Immersion Ion Implantation and Deposition Using Vacuum Arc Plasma Sources. *J. Vac. Sci. Technol. B Microelectron. Nanometer Struct. Process. Meas. Phenom.* **1994**, *12* (2), 815–820. https://doi.org/10.1116/1.587351.

65. Wang, C. Localized Surface Functionalization with Atmospheric-Pressure Microplasma Jet for Cell-on-a-Chip Applications. Doctor of Philosophy, Drexel University, Philadelphia, Pennsylvania, United States, 2016. https://doi.org/10.17918/etd-7076.

66. Souza, S. T.; Agra, L. C.; Santos, C. E. A.; Barreto, E.; Hickmann, J. M.; Fonseca, E. J. S. Macrophage Adhesion on Fibronectin Evokes an Increase in the Elastic Property of the Cell Membrane and Cytoskeleton: An Atomic Force Microscopy Study. *Eur. Biophys. J.* **2014**, *43* (12), 573–579. https://doi.org/10.1007/s00249-014-0988-3.

67. Parsey, G.; Lietz, A. M.; Kushner, M. J. Guided Plasma Jets Directed onto Wet Surfaces: Angular Dependence and Control. *J. Phys. Appl. Phys.* **2021**, *54* (4), 045206. https://doi.org/10.1088/1361-6463/abbf1a.

68. Tang, R.; Etzion, Y. Comparative Studies on the Water Evaporation Rate from a Wetted Surface and That from a Free Water Surface. *Build. Environ.* **2004**, *39* (1), 77–86. https://doi.org/10.1016/j.buildenv.2003.07.007.

69. Meyyappan, M.; Delzeit, L.; Cassell, A.; Hash, D. Carbon Nanotube Growth by PECVD: A Review. *Plasma Sources Sci. Technol.* **2003**, *12* (2), 205–216. https://doi.org/10.1088/0963-0252/12/2/312.

70. Kushner, M. J. A Model for the Discharge Kinetics and Plasma Chemistry during Plasma Enhanced Chemical Vapor Deposition of Amorphous Silicon. *J. Appl. Phys.* **1988**, *63* (8), 2532–2551. https://doi.org/10.1063/1.340989.

71. Ferreira, A.; Roussy, A.; Kernaflen, C.; Gleispach, D.; Hayderer, G.; Gris, H.; Besnard, J. Virtual Metrology Models for Predicting Avera PECVD Oxide Film Thickne. In *2011 IEEE/SEMI Advanced Semiconductor Manufacturing Conference*; IEEE: Saratoga Springs, NY, USA, 2011; pp. 1–6. https://doi.org/10.1109/ASMC.2011.5898187.

72. Kabongo, G. L.; Mothudi, B. M.; Dhlamini, M. S. Advanced Development of Sustainable PECVD Semitransparent Photovoltaics: A Review. *Front. Mater.* **2021**, *8*, 762030. https://doi.org/10.3389/fmats.2021.762030.

73. Huang, H.-J.; Kau, L.-H.; Wang, H.-S.; Hsieh, Y.-L.; Lee, C.-C.; Fuh, Y.-K.; Li, T. T. Large-Scale Data Analysis of PECVD Amorphous Silicon Interface Passivation Layer via the Optical Emission Spectra for Parameterized PCA. *Int. J. Adv. Manuf. Technol.* **2019**, *101* (1–4), 329–337. https://doi.org/10.1007/s00170-018-2938-1.

74. Roca I Cabarrocas, P.; Fontcuberta I Morral, A.; Kalache, B.; Kasouit, S. Microcrystalline Silicon Thin-Films Grown by Plasma Enhanced Chemical Vapour Deposition: Growth Mechanisms and Grain Size Control. *Solid State Phenom.* **2003**, *93*, 257–268. https://doi.org/10.4028/www.scientific.net/SSP.93.257.

75. Hegedüs, N.; Balázsi, K.; Balázsi, C. Silicon Nitride and Hydrogenated Silicon Nitride Thin Films: A Review of Fabrication Methods and Applications. *Materials* **2021**, *14* (19), 5658. https://doi.org/10.3390/ma14195658.

76. Iliescu, C. A Comprehensive Review on Thin Film Depositions on PECVD Reactors. *Ann. Acad. Romanian Sci. Ser. Sci. Technol. Inf.* **2021**, *14* (1–2), 12–24. https://doi.org/10.56082/annalsarsciinfo.2021.1-2.12.

77. Choi, E. H. *Plasma Biosciences and Medicine.* Part of the book series "Topics in Applied Physics"; Springer Nature Singapore: Singapore, 2023; Vol. 148. https://doi.org/10.1007/978-981-19-7935-4.

78. Von Woedtke, Th.; Reuter, S.; Masur, K.; Weltmann, K.-D. Plasmas for Medicine. *Phys. Rep.* **2013**, *530* (4), 291–320. https://doi.org/10.1016/j.physrep.2013.05.005.

79. Keidar, M. *Plasma Cancer Therapy.* Springer Series on Atomic, Optical, and Plasma Physics; Springer International Publishing: Cham, 2020; Vol. 115. https://doi.org/10.1007/978-3-030-49966-2.

80. Laroussi, M.; Bekeschus, S.; Keidar, M.; Bogaerts, A.; Fridman, A.; Lu, X.; Ostrikov, K.; Hori, M.; Stapelmann, K.; Miller, V.; Reuter, S.; Laux, C.; Mesbah, A.; Walsh, J.; Jiang, C.; Thagard, S. M.; Tanaka, H.; Liu, D.; Yan, D.; Yusupov, M. Low-Temperature Plasma for Biology, Hygiene, and Medicine: Perspective and Roadmap. *IEEE Trans. Radiat. Plasma Med. Sci.* **2022**, *6* (2), 127–157. https://doi.org/10.1109/TRPMS.2021.3135118.

81. Adamovich, I.; Agarwal, S.; Ahedo, E.; Alves, L. L.; Baalrud, S.; Babaeva, N.; Bogaerts, A.; Bourdon, A.; Bruggeman, P. J.; Canal, C.; Choi, E. H.; Coulombe, S.; Donkó, Z.; Graves, D. B.; Hamaguchi, S.; Hegemann, D.; Hori, M.; Kim, H.-H.; Kroesen, G. M. W.; Kushner, M. J.; Laricchiuta, A.; Li, X.; Magin, T. E.; Mededovic Thagard, S.; Miller, V.; Murphy, A. B.; Oehrlein, G. S.; Puac, N.; Sankaran, R. M.; Samukawa, S.; Shiratani, M.; Šimek, M.; Tarasenko, N.; Terashima, K.; Thomas Jr, E.; Trieschmann, J.; Tsikata, S.; Turner, M. M.; Van Der Walt, I. J.; Van De Sanden, M. C. M.; Von Woedtke, T. The 2022 Plasma Roadmap: Low Temperature Plasma Science and Technology. *J. Phys. Appl. Phys.* **2022**, *55* (37), 373001. https://doi.org/10.1088/1361-6463/ac5e1c.

82. Graves, D. B. The Emerging Role of Reactive Oxygen and Nitrogen Species in Redox Biology and Some Implications for Plasma Applications to Medicine and Biology. *J. Phys. Appl. Phys.* **2012**, *45* (26), 263001. https://doi.org/10.1088/0022-3727/45/26/263001.

83. Graves, D. B. Reactive Species from Cold Atmospheric Plasma: Implications for Cancer Therapy. *Plasma Process. Polym.* **2014**, *11* (12), 1120–1127. https://doi.org/10.1002/ppap.201400068.

84. Yan, D.; Talbot, A.; Nourmohammadi, N.; Cheng, X.; Canady, J.; Sherman, J.; Keidar, M. Principles of Using Cold Atmospheric Plasma Stimulated Media for Cancer Treatment. *Sci. Rep.* **2015**, *5* (1), 18339. https://doi.org/10.1038/srep18339.

85. Keidar, M.; Yan, D.; Beilis, I. I.; Trink, B.; Sherman, J. H. Plasmas for Treating Cancer: Opportunities for Adaptive and Self-Adaptive Approaches. *Trends Biotechnol.* **2018**, *36* (6), 586–593. https://doi.org/10.1016/j.tibtech.2017.06.013.

86. Keidar, M. A Prospectus on Innovations in the Plasma Treatment of Cancer. *Phys. Plasmas* **2018**, *25* (8), 083504. https://doi.org/10.1063/1.5034355.

87. Cheng, X.; Murphy, W.; Recek, N.; Yan, D.; Cvelbar, U.; Vesel, A.; Mozetič, M.; Canady, J.; Keidar, M.; Sherman, J. H. Synergistic Effect of Gold Nanoparticles and Cold Plasma on Glioblastoma Cancer Therapy. *J. Phys. Appl. Phys.* **2014**, *47* (33), 335402. https://doi.org/10.1088/0022-3727/47/33/335402.

88. Keidar, M. Plasma for Cancer Treatment. *Plasma Sources Sci. Technol.* **2015**, *24* (3), 033001. https://doi.org/10.1088/0963-0252/24/3/033001.

89. Bisag, A.; Manzini, M.; Simoncelli, E.; Stancampiano, A.; Tonini, R.; Gherardi, M.; Colombo, V. Cold Atmospheric Pressure Plasma Treatment to Assist the Restoration of the Apical Region of a Root Canal in Endodontic Procedures. *Clin. Plasma Med.* **2020**, *19–20*, 100100. https://doi.org/10.1016/j.cpme.2020.100100.

90. Lu, X.; Liu, D.; Xian, Y.; Nie, L.; Cao, Y.; He, G. Cold Atmospheric-Pressure Air Plasma Jet: Physics and Opportunities. *Phys. Plasmas* **2021**, *28* (10), 100501. https://doi.org/10.1063/5.0067478.

91. Lin, L.; Lyu, Y.; Trink, B.; Canady, J.; Keidar, M. Cold Atmospheric Helium Plasma Jet in Humid Air Environment. *J. Appl. Phys.* **2019**, *125* (15), 153301. https://doi.org/10.1063/1.5086177.

92. Lin, L.; Keidar, M. A Map of Control for Cold Atmospheric Plasma Jets: From Physical Mechanisms to Optimizations. *Appl. Phys. Rev.* **2021**, *8* (1), 011306. https://doi.org/10.1063/5.0022534.

93. Ma, Y.; Ha, C. S.; Hwang, S. W.; Lee, H. J.; Kim, G. C.; Lee, K.-W.; Song, K. Non-Thermal Atmospheric Pressure Plasma Preferentially Induces Apoptosis in P53-Mutated Cancer Cells by Activating ROS Stress-Response Pathways. *PLoS ONE* **2014**, *9* (4), e91947. https://doi.org/10.1371/journal.pone.0091947.

94. Uchida, G.; Nakajima, A.; Ito, T.; Takenaka, K.; Kawasaki, T.; Koga, K.; Shiratani, M.; Setsuhara, Y. Effects of Nonthermal Plasma Jet Irradiation on the Selective Production of H_2O_2 and NO_2^- in Liquid Water. *J. Appl. Phys.* **2016**, *120* (20), 203302. https://doi.org/10.1063/1.4968568.

95. Yan, D.; Horkowitz, A.; Wang, Q.; Keidar, M. On the Selective Killing of Cold Atmospheric Plasma Cancer Treatment: Status and Beyond. *Plasma Process. Polym.* **2021**, *18* (10), 2100020. https://doi.org/10.1002/ppap.202100020.

96. Anikin, N. B.; Starikovskaia, S. M.; Starikovskii, A. Y. Oxidation of Saturated Hydrocarbons under the Effect of Nanosecond Pulsed

Space Discharge. *J. Phys. Appl. Phys.* **2006**, *39* (15), 3244–3252. https://doi.org/10.1088/0022-3727/39/15/006.

97. Yan, D.; Lin, L.; Sherman, J. H.; Canady, J.; Trink, B.; Keidar, M. The Correlation Between the Cytotoxicity of Cold Atmospheric Plasma and the Extracellular H_2O_2 Scavenging Rate. *IEEE Trans. Radiat. Plasma Med. Sci.* **2018**, *2* (6), 618–623. https://doi.org/10.1109/TRPMS.2018.2871134.

98. Yan, D.; Lin, L.; Xu, W.; Nourmohammadi, N.; Sherman, J. H.; Keidar, M. Cold Plasma-Based Control of the Activation of Pancreatic Adenocarcinoma Cells. *J. Phys. Appl. Phys.* **2019**, *52* (44), 445202. https://doi.org/10.1088/1361-6463/ab36d4.

99. Wang, M.; Holmes, B.; Cheng, X.; Zhu, W.; Keidar, M.; Zhang, L. G. Cold Atmospheric Plasma for Selectively Ablating Metastatic Breast Cancer Cells. *PLoS ONE* **2013**, *8* (9), e73741. https://doi.org/10.1371/journal.pone.0073741.

100. Uchida, G.; Takenaka, K.; Takeda, K.; Ishikawa, K.; Hori, M.; Setsuhara, Y. Selective Production of Reactive Oxygen and Nitrogen Species in the Plasma-Treated Water by Using a Nonthermal High-Frequency Plasma Jet. *Jpn. J. Appl. Phys.* **2018**, *57* (1), 0102B4. https://doi.org/10.7567/JJAP.57.0102B4.

101. Cheng, X.; Rajjoub, K.; Sherman, J.; Canady, J.; Recek, N.; Yan, D.; Bian, K.; Murad, F.; Keidar, M. Cold Plasma Accelerates the Uptake of Gold Nanoparticles Into Glioblastoma Cells. *Plasma Process. Polym.* **2015**, *12* (12), 1364–1369. https://doi.org/10.1002/ppap.201500093.

102. Baik, K. Y.; Huh, Y. H.; Kim, Y. H.; Kim, J.; Kim, M. S.; Park, H.-K.; Choi, E. H.; Park, B. The Role of Free Radicals in Hemolytic Toxicity Induced by Atmospheric-Pressure Plasma Jet. *Oxid. Med. Cell. Longev.* **2017**, *2017* (1), 1289041. https://doi.org/10.1155/2017/1289041.

103. Hirst, A. M.; Frame, F. M.; Arya, M.; Maitland, N. J.; O'Connell, D. Low Temperature Plasmas as Emerging Cancer Therapeutics: The State of Play and Thoughts for the Future. *Tumor Biol.* **2016**, *37* (6), 7021–7031. https://doi.org/10.1007/s13277-016-4911-7.

104. Liu, J.-R.; Wu, Y.-M.; Xu, G.-M.; Gao, L.-G.; Ma, Y.; Shi, X.-M.; Zhang, G.-J. Low-Temperature Plasma Induced Melanoma Apoptosis by Triggering a P53/PIGs/Caspase-Dependent Pathway *in Vivo* and *in Vitro*. *J. Phys. Appl. Phys.* **2019**, *52* (31), 315204. https://doi.org/10.1088/1361-6463/ab1ca9.

105. Liu, Z.; Xu, D.; Liu, D.; Cui, Q.; Cai, H.; Li, Q.; Chen, H.; Kong, M. G. Production of Simplex RNS and ROS by Nanosecond Pulse N_2/

O_2 Plasma Jets with Homogeneous Shielding Gas for Inducing Myeloma Cell Apoptosis. *J. Phys. Appl. Phys.* **2017**, *50* (19), 195204. https://doi.org/10.1088/1361-6463/aa66f0.

106. Soni, V.; Adhikari, M.; Simonyan, H.; Lin, L.; Sherman, J. H.; Young, C. N.; Keidar, M. In Vitro and In Vivo Enhancement of Temozolomide Effect in Human Glioblastoma by Non-Invasive Application of Cold Atmospheric Plasma. *Cancers* **2021**, *13* (17), 4485. https://doi.org/10.3390/cancers13174485.

107. Soni, V.; Adhikari, M.; Lin, L.; Sherman, J. H.; Keidar, M. Theranostic Potential of Adaptive Cold Atmospheric Plasma with Temozolomide to Checkmate Glioblastoma: An In Vitro Study. *Cancers* **2022**, *14* (13), 3116. https://doi.org/10.3390/cancers14133116.

108. Soni, V.; Dawson, T.; Lin, L.; Crandall, K.; Sherman, J.; Keidar, M. High-Throughput RNA-Seq and In-Silico Analysis of Glioblastoma Cells Treated with Cold Atmospheric Plasma and Temozolomide. June 20, 2024. https://doi.org/10.21203/rs.3.rs-4569010/v1.

109. Schweigert, I. V.; Zakrevsky, D. E.; Gugin, P. P.; Milakhina, E. V.; Biryukov, M. M.; Keidar, M.; Koval, O. A. Effect of Voltage Pulse Duration on Electrophysical and Thermal Characteristics of Cold Atmospheric Plasma Jet. *Plasma Sources Sci. Technol.* **2022**, *31* (11), 114004. https://doi.org/10.1088/1361-6595/aca120.

110. Yan, D.; Wang, Q.; Adhikari, M.; Malyavko, A.; Lin, L.; Zolotukhin, D. B.; Yao, X.; Kirschner, M.; Sherman, J. H.; Keidar, M. A Physically Triggered Cell Death via Transbarrier Cold Atmospheric Plasma Cancer Treatment. *ACS Appl. Mater. Interfaces* **2020**, *12* (31), 34548–34563. https://doi.org/10.1021/acsami.0c06500.

111. Yao, X.; Yan, D.; Lin, L.; Sherman, J. H.; Peters, K. B.; Keir, S. T.; Keidar, M. Cold Plasma Discharge Tube Enhances Antitumoral Efficacy of Temozolomide. *ACS Appl. Bio Mater.* **2022**, *5* (4), 1610–1623. https://doi.org/10.1021/acsabm.2c00018.

112. Lin, L.; McCraw, M. R.; Uluutku, B.; Liu, Y.; Yan, D.; Soni, V.; Horkowitz, A.; Yao, X.; Limanowski, R.; Solares, S. D.; Beilis, I. I.; Keidar, M. Cell Membrane Oscillations under Radiofrequency Electromagnetic Modulation. *Langmuir* **2023**, *39* (9), 3320–3331. https://doi.org/10.1021/acs.langmuir.2c03181.

113. Boeuf, J.-P.; Yang, L. L.; Pitchford, L. C. Dynamics of a Guided Streamer ('Plasma Bullet') in a Helium Jet in Air at Atmospheric Pressure. *J. Phys. Appl. Phys.* **2013**, *46* (1), 015201. https://doi.org/10.1088/0022-3727/46/1/015201.

114. Sigmond, R. S. The Residual Streamer Channel: Return Strokes and Secondary Streamers. *J. Appl. Phys.* **1984**, *56* (5), 1355–1370. https://doi.org/10.1063/1.334126.

115. Schweigert, I. V.; Vagapov, S.; Lin, L.; Keidar, M. Increase of Electric Field on Streamer/Surfaces Contact Area with Using External Electrode. *J. Phys. Conf. Ser.* **2019**, *1394* (1), 012005. https://doi.org/10.1088/1742-6596/1394/1/012005.

116. Lu, X.; Naidis, G. V.; Laroussi, M.; Ostrikov, K. Guided Ionization Waves: Theory and Experiments. *Phys. Rep.* **2014**, *540* (3), 123–166. https://doi.org/10.1016/j.physrep.2014.02.006.

117. Lu, X.; Ostrikov, K. (Ken). Guided Ionization Waves: The Physics of Repeatability. *Appl. Phys. Rev.* **2018**, *5* (3), 031102. https://doi.org/10.1063/1.5031445.

118. Lietz, A. M.; Barnat, E. V.; Foster, J. E.; Kushner, M. J. Ionization Wave Propagation in a He Plasma Jet in a Controlled Gas Environment. *J. Appl. Phys.* **2020**, *128* (8), 083301. https://doi.org/10.1063/5.0020264.

119. Babaeva, N. Y.; Kushner, M. J. Self-Organization of Single Filaments and Diffusive Plasmas during a Single Pulse in Dielectric-Barrier Discharges. *Plasma Sources Sci. Technol.* **2014**, *23* (6), 065047. https://doi.org/10.1088/0963-0252/23/6/065047.

120. Schoenbach, K. H.; Moselhy, M.; Shi, W. Self-Organization in Cathode Boundary Layer Microdischarges. *Plasma Sources Sci. Technol.* **2004**, *13* (1), 177–185. https://doi.org/10.1088/0963-0252/13/1/023.

121. Foster, J. E.; Kovach, Y. E.; Lai, J.; Garcia, M. C. Self-Organization in 1 Atm DC Glows with Liquid Anodes: Current Understanding and Potential Applications. *Plasma Sources Sci. Technol.* **2020**, *29* (3), 034004. https://doi.org/10.1088/1361-6595/ab7089.

122. Konina, K.; Raskar, S.; Adamovich, I. V.; Kushner, M. J. Atmospheric Pressure Plasmas Interacting with Wet and Dry Microchannels: Reverse Surface Ionization Waves. *Plasma Sources Sci. Technol.* **2024**, *33* (1), 015002. https://doi.org/10.1088/1361-6595/ad171c.

123. Konina, K.; Kruszelnicki, J.; Meyer, M. E.; Kushner, M. J. Surface Ionization Waves Propagating over Non-Planar Substrates: Wavy Surfaces, Cut-Pores and Droplets. *Plasma Sources Sci. Technol.* **2022**, *31* (11), 115001. https://doi.org/10.1088/1361-6595/ac9a6c.

124. Schweigert, I. V.; Vagapov, S.; Lin, L.; Keidar, M. Enhancement of Atmospheric Plasma Jet–Target Interaction with an External Ring

Electrode. *J. Phys. Appl. Phys.* **2019**, *52* (29), 295201. https://doi.org/10.1088/1361-6463/ab1319.

125. Moiseev, T.; Misra, N. N.; Patil, S.; Cullen, P. J.; Bourke, P.; Keener, K. M.; Mosnier, J. P. Post-Discharge Gas Composition of a Large-Gap DBD in Humid Air by UV–Vis Absorption Spectroscopy. *Plasma Sources Sci. Technol.* **2014**, *23* (6), 065033. https://doi.org/10.1088/0963-0252/23/6/065033.

126. Mohd Ridzuan Tan, N. D.; Abdul Munir, F.; Mohd Tahir, M.; Saputro, H.; Mikami, M. Preliminary Investigation of Using DBD Plasma for Application in Micro Combustors. *J. Adv. Res. Fluid Mech. Therm. Sci.* **2021**, *82* (1), 105–112. https://doi.org/10.37934/arfmts.82.1.105112.

127. Foster, J. E.; Weatherford, B.; Gillman, E.; Yee, B. Underwater Operation of a DBD Plasma Jet. *Plasma Sources Sci. Technol.* **2010**, *19* (2), 025001. https://doi.org/10.1088/0963-0252/19/2/025001.

128. Biset-Peiró, M.; Guilera, J.; Andreu, T. Ignition of CO_2 Methanation Using DBD-Plasma Catalysis in an Adiabatic Reactor. *Chem. Eng. J.* **2022**, *433*, 133638. https://doi.org/10.1016/j.cej.2021.133638.

129. Abolmasov, S. N.; Tachibana, K.; Shirafuji, T. Mechanisms of Pattern Formation in Dielectric Barrier Discharges. *IEEE Trans. Plasma Sci.* **2011**, *39* (11), 2090–2091. https://doi.org/10.1109/TPS.2011.2160362.

130. Chirokov, A.; Gutsol, A.; Fridman, A.; Sieber, K. D.; Grace, J. M.; Robinson, K. S. A Study of Two-Dimensional Microdischarge Pattern Formation in Dielectric Barrier Discharges. *Plasma Chem. Plasma Process.* **2006**, *26* (2), 127–135. https://doi.org/10.1007/s11090-006-9007-5.

131. Shi, J.; Zhong, F.; Zhang, J.; Liu, D. W.; Kong, M. G. A Hypersonic Plasma Bullet Train Traveling in an Atmospheric Dielectric-Barrier Discharge Jet. *Phys. Plasmas* **2008**, *15* (1), 013504. https://doi.org/10.1063/1.2828551.

132. Sharma, N. K.; Misra, S.; Varun; Pal, U. N. Experimental and Simulation Analysis of Dielectric Barrier Discharge Based Pulsed Cold Atmospheric Pressure Plasma Jet. *Phys. Plasmas* **2020**, *27* (11), 113502. https://doi.org/10.1063/5.0018901.

Chapter 2

Plasma Diagnostics and Controls Using Machine Learning

Li Lin, Anmol Taploo, Vikas Soni, and Michael Keidar

School of Engineering and Applied Science, The George Washington University, Washington, DC 20052, USA

lilin@email.gwu.edu

2.1 The Interactions of Plasma Parameters

As introduced in the previous chapter, plasma properties can be computed by Maxwell's equations and the conservation laws. These equations describe the relationship between these parameters. A flow chart is shown in Fig. 2.1. The chart contains blocks and arrows while the blocks represent the physical parameters, the arrows are mostly the ways of processing the parameters such as the equations and experimental hardware. The dashed-line arrows represent the signal sent and received. What the users can usually control is the energy input such as the external E × B field for the inductively coupled plasma (ICP) and thermal energy that can be conducted by heating and laser beams. The discharge waveform applied through the electrodes of a capacitively coupled plasma (CCP) is the boundary conditions of Maxwell's

Machine Learning Plasmas and the Neuromorphic Plasma Chemistry
Edited by Li Lin and Michael Keidar

ISBN 978-981-5129-84-7 (Hardcover), 978-1-003-71013-4 (eBook)
www.jennystanford.com

DOI: 10.1201/9781003710134-2

equations. Combined with the charge distributions in space, such as the free electrons and ions as a part of the chemical composition, the drift velocities of these charges are determined at that moment. Then, combined with the momentum transfers of particle collisions, the particles' kinetic energies are determined. Some inelastic collisions can also modify the species' energies such as thermal energy release through combustion, which can be quantified by computing the standard enthalpy of formation. The values can be described as the energy distribution functions which also represent the temperature of these species. For a thermal non-equilibrium plasma, different species may have different temperatures. Their kinetic energy, together with the cross sections, determines the rate coefficients of each different type of collisions, following Eq. 1.2 in Chapter 1, and the finally the rate equation plays as the net source of species in the mass equation to determine the chemical composition in the next moment. On the other hand, the users can monitor the species concentrations using spectroscopies such as mass spectroscopy (MS), laser induced fluorescence (LIF), Fourier-transform infrared spectroscopy (FTIR), etc.[1–5] Also, beam scattering methods such as laser Thomson scattering (LTS) and Rayleigh microwave scattering (RMS, for small-size plasma) can be used to measure the free electron density.[6–11] Please note that the commonly used optical emission spectroscopy (OES) does not directly reflect any species concentrations, especially those at the ground states. OES signals can provide the number of photons at each wavelength but self-absorption of optical emissions in the thick plasmas must be considered. On the other hand, the users can also measure the temperatures, or the kinetic energy distribution functions, of the species as shown in Fig. 2.1.

Overall, one can see that a clear control and diagnostics loop from the user to their control parameters of species flow and finally be monitored. This is a typical fluid mechanics control loop. When there is chemistry involved but no plasma contained, the map shown in Fig. 2.1 will expand to the collisions. The types of collisions can be found in Fig. 1.1, Chapter 1. Many of the inelastic collisions can release or absorb energy, such as the redox reactions in combustions can release heat. Their exact values can be quantified by computing the variation of standard enthalpy of formation. The reactant temperature can thus be modified and so will the reaction rate coefficients in the next moment. This is the physical mechanisms of how a chemical reaction wave propagates in a medium at stationery, such as the expansion of flame front, while in a combustion chamber, the gas-phase medium also has a bulk velocity. The temperature can also be measured by means of a temperature sensor, etc. while the reaction products can be measured using spectroscopies etc. This completes a control-diagnostics cycle of chemistry-coupled fluid mechanics.

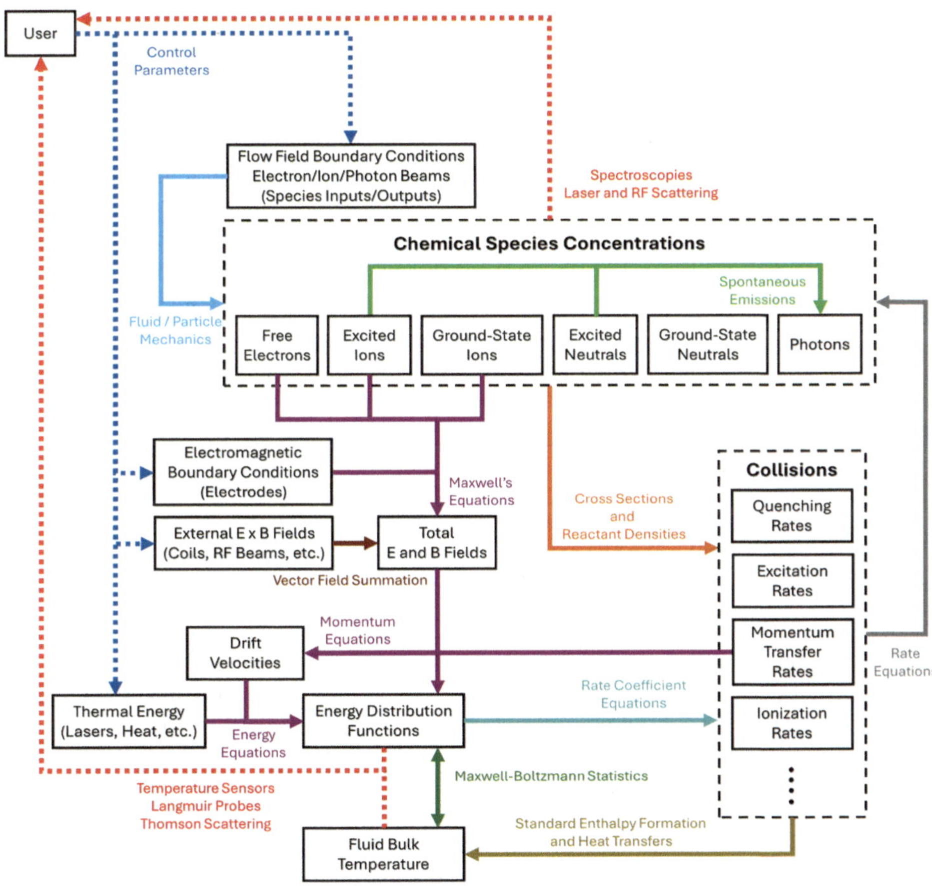

Figure 2.1 The flow chart of plasma parameters, control parameters, and diagnostics information.

When plasmas are added, the map is expanded to the entire Fig. 2.1. To control and diagnostics, there are not too many other methods that can be used on directly measuring the drift velocity fields of free electrons and ions, and their own E × B fields excluding the external fields. However, there are many plasmas spatialized diagnostics that can be used to measure the EEDF, ion density, etc.

Therefore, like other close-loop controls, i.e., feed-back controls, the plasma control loop also follows the basic schematics as shown in Fig. 2.2, although the parameter interactions shown in Fig. 2.1 are complicated. The input of plasma system is the control parameter which is a vector $\boldsymbol{x}(t)$ and the plasma parameters is also a list, a.k.a. a vector, denoted as $\boldsymbol{y}(t)$ because in general, we must consider the plasma as a multi-input-multi-output system (MIMO). The plasma system is thus

$$\boldsymbol{y}(t) = f[\boldsymbol{x}(t)] \tag{2.1}$$

where the function f is the plasma physics and chemistry which is determined by our nature, and t is time. The diagnostics systems acquire $\boldsymbol{y}(t)$ and send it to the controller.

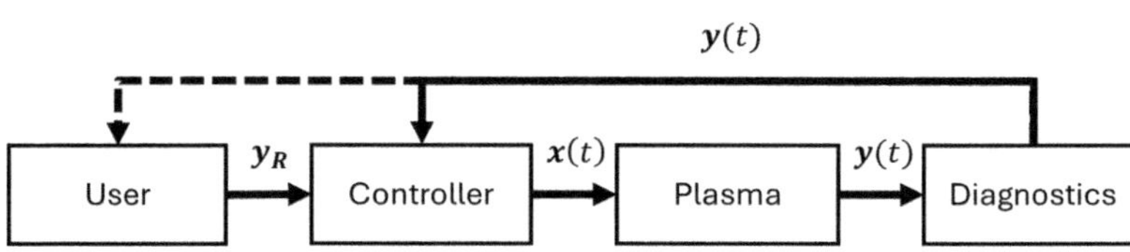

Figure 2.2 The control loop of plasmas.

The controller needs to determine what will be the best $\boldsymbol{x}(t)$ so that the resulting $\boldsymbol{y}(t)$ can approach to the user's requirement $\boldsymbol{y}_R$. Mathematically, the algorithm in the controller can be represented as the following based on the control theory:

$$\boldsymbol{x}_{\mathbf{opt}}(t) = \arg\min_{\boldsymbol{x}(t)} [\epsilon(\boldsymbol{x})] \tag{2.2a}$$

$$\epsilon(\boldsymbol{x}) = \Sigma|\boldsymbol{y}_R - \boldsymbol{y}| = \Sigma|\boldsymbol{y}_R - f(\boldsymbol{x})| \tag{2.2b}$$

where $\boldsymbol{x}_{\text{opt}}$ is the optimized $\boldsymbol{x}$, ϵ is the error which is defined as the difference between $\boldsymbol{y}$ and the user required one $\boldsymbol{y}_R$. Equation (2.2b) is also known as "loss function". Please note that in Eq. (2.2b), the error is defined as the sum of absolute difference between $\boldsymbol{y}_R$ and $\boldsymbol{y}$. In many applications, a more commonly seen definition is

$$\epsilon(\boldsymbol{x}) = \Sigma[(\boldsymbol{y}_R - \boldsymbol{y})^2] \tag{2.3}$$

However, due to the complexity of the plasma parameters as shown in Fig. 2.1, it is usually challenging to control using regular controller algorithm such as the proportional–integral–derivative (PID) control. However, controlling the plasma generator using machine learning (ML) provides a solution, which will be introduced in the next section.

2.2 Fitting with Machine Learning

ML is a subset of artificial intelligence (AI) that focuses on developing algorithms that allow computers to learn patterns from data and make decisions or predictions without being explicitly programmed for specific tasks. The core idea is to enable a system to improve its performance on a task by learning from experience (data). Mathematically, machine learning can be viewed as a function approximation problem, where a function $f: \mathbb{R}^n \rightarrow \mathbb{R}^m$ maps input data $\boldsymbol{x} \in \mathbb{R}^n$ to output data $\boldsymbol{y} \in \mathbb{R}^m$. At the heart of ML is the concept of a model, typically represented by a parameterized function $f(\boldsymbol{x}; \theta)$, where θ is a set of parameters. The goal of ML is to find the optimal parameters θ_{opt} that minimize a loss function Eq. (2.2b) or Eq. (2.3). This optimization is often performed using gradient-based methods, where the gradient of the loss function with respect to the parameters is computed and used to update the parameters iteratively. For example, for the scenario of autonomous driving, the input will be the real-time image data of a front camera on the vehicle and the output will be the control parameters such as steering, brake, etc.[12,13]

To introduce the idea in detail, let's consider a simple relationship between the input x and the required output $\boldsymbol{y_R}$ is known to be a quadratic function:

$$y_R = 3x \tag{2.4}$$

Equation (2.4) is the goal of a ML user to find. In other words, the user must quantify a mathematical model that can output the correct result y based on any input x. To find such a relationship, one can first design a polynomial model:

$$y = \theta x \tag{2.5}$$

Therefore, the error is defined by the loss function as

$$\epsilon = \Sigma_i \left(y_{R,i} - \theta x_i\right)^2 \tag{2.6}$$

Using the gradient descent method to train the model Eq. (2.5), one needs to find the derivative of the loss function with respect to the parameter θ that

$$\nabla\epsilon = -2\Sigma_i \left(y_{R,i} x_i - \theta x_i^2\right) \tag{2.7}$$

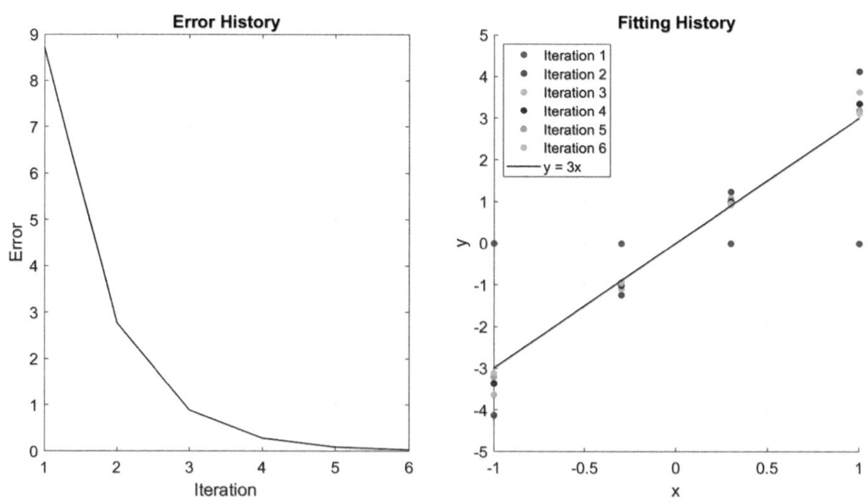

Figure 2.3 The resulting ***y*** values based on the initial guess of ***x*** values versus the required ***y*** values $\boldsymbol{y_R}$.

Then, update the parameter θ as

$$\theta \leftarrow \theta - \eta \nabla \epsilon \tag{2.8}$$

where η is the learning rate which is a positive value small enough to ensure the stability of the fitting process. In this example, we choose $\eta = 0.1$. Next, select randomly of x values, for example, the $\boldsymbol{x}$ vector can be [−1, −0.3, 0.3, 1], and the initial guess θ can be 5. The training history of model Eq. (2.5) is thus shown in Fig. 2.3. One can see that after 6 iterations, the total error is approaching zero, and the resulting $\boldsymbol{y}$ in light blue is very close to $\boldsymbol{y_R}$ as shown in the black line, with the given $\boldsymbol{x}$. The resulting θ after 6 iterations is 3.155, very close to 3 as required in Eq. 2.4.

However, in cases where the system is too complex to be described by a traditional, such as a plasma system as shown in Fig. 2.1, explicit model (e.g., a highly nonlinear system with numerous interacting variables), a neural network can be used to approximate the underlying function $f(\boldsymbol{x};\ \theta)$. An artificial neural network (ANN) acts as a universal function approximator, meaning it can learn to represent the mapping even when the exact form of the function is unknown. The neural network is designed with layers of neurons, where each neuron is a variable, and a layer of neurons means a vector, denoted as $\boldsymbol{h}^{(l)}$ where the superscript l means the lth layer. Therefore, the input of an ANN is

$$\boldsymbol{x} = \boldsymbol{h}^{(l=1)} \tag{2.9a}$$

Each neuron is a result of an activation function σ and the input of such a function is a weighed-biased linear combination of all the neurons in the previous layer that

$$\boldsymbol{h}^{(l)} = \sigma[\boldsymbol{w}^{(l)}\boldsymbol{h}^{(l-1)} + \boldsymbol{b}^{(l)}] \tag{2.9b}$$

where $\boldsymbol{w}$ is a set of matrices of weights and $\boldsymbol{b}$ is a set of matrices of bias. Thus, a common design of the network's parameters is $\theta = \{\boldsymbol{w},\ \boldsymbol{b}\}$. Commonly used activation functions including Sigmoid, ReLU, etc. will output a significant value only when they receive an input over a threshold. Table 2.1 shows some commonly used activation functions summarized. We will have more discussions and figures of these functions in Chapter 4. However, they also have different behaviors, such as the Sigmoid function converges at high input, and they surely significantly determine the overall performance of an ANN, such as the error gradient sensitivity, etc. which will be discussed in the next section. At the last layer, the Lth layer, the output of ANN can be mathematically described as

$$\boldsymbol{y} = f(\boldsymbol{x};\theta) = \boldsymbol{h}^{(L)} = \boldsymbol{w}^{(L)}\,\boldsymbol{h}^{(L-1)} + \boldsymbol{b}^{(L)} \tag{2.9c}$$

Table 2.1 Activation functions

Name	Mathematical Expression
Sigmoid	$\frac{1}{1+\exp(-x)}$
ReLU	$\begin{cases} 0 \text{ if } x<0 \\ x \text{ if } x\geq 0 \end{cases}$
tanh	$\frac{\exp x-\exp(-x)}{\exp x+\exp(-x)}$
Softplus	$\ln(1+\exp x)$

The schematic of a fully connected ANN is shown in Fig. 2.4. For example, the value of neuron (2, 2) can be computed according to Eq. (2.9b) as $\sigma(w_{1,1\to2,2}n_{1,1} + b_{1,1\to2,2} + w_{1,2\to2,2}n_{1,2} + b_{1,2\to2,2} + w_{1,x\to2,2}n_{1,x} + b_{1,x\to2,2}$. Please note that there are many commonly used techniques on the input and output of each layer such as L2 regularization and normalization, which will apply more hyperparameters in addition to the weights and bias.

However, while fully connected networks, where every neuron in one layer is connected to every neuron in the subsequent layer, are powerful and versatile, they can be inefficient and prone to overfitting for certain tasks, particularly when dealing with high-dimensional data like images, sequences, or structured inputs.

To address these challenges, different network topologies have been developed, including Convolutional Neural Networks (CNNs), Recurrent Neural Networks (RNNs), and others.

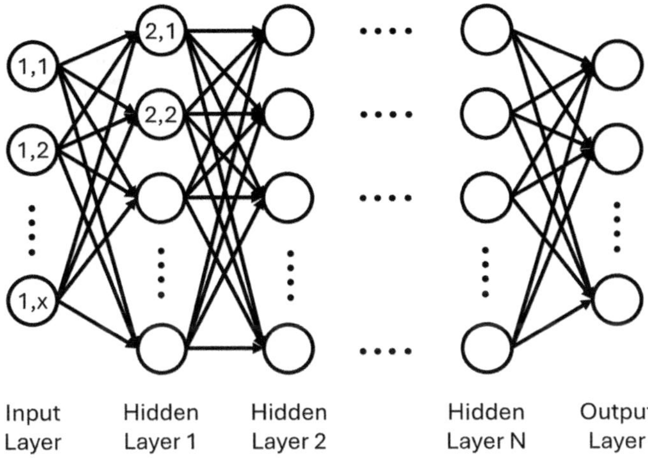

Figure 2.4 The schematic of a fully connected ANN.

CNN contains convolution layers and pooling layers before its fully connected layers. The convolution layers apply convolutional filters (kernels) to the input data, which helps to capture local patterns such as edges, textures, and shapes. A convolution operation slides a filter over the input data and produces a feature map, highlighting the presence of specific patterns at various locations. Mathematically, the convolution operation for a 2D input $\boldsymbol{x}$ and a filter $\boldsymbol{K}$ is expressed as:

$$\boldsymbol{y}(i,j) = \Sigma_m \Sigma_n \boldsymbol{x}(i+m, j+n) \cdot \boldsymbol{K}(m,n) \tag{2.10}$$

where (i, j) is the position on the 2D input. Pooling layers reduce the spatial dimensions of the feature maps, helping to reduce the computational load and the risk of overfitting. Common pooling operations include max pooling, which takes the maximum value within a window, and average pooling, which computes the average. For example, max pooling is one of the most common types of pooling operations. In max pooling, the feature map is divided into non-overlapping (or sometimes overlapping) rectangular regions, and the maximum value within each region is selected to represent the region in the output feature map.

RNNs are designed to handle sequential data, where the order of inputs matters. RNNs are particularly effective for tasks involving time series, natural language processing, and any data that can be represented as a sequence. RNN is more like the structure of CPN which will be discussed in Chapter 4. Unlike fully connected networks, RNNs have connections that loop back, allowing information to persist over time. This means that the output from a neuron at one time step is used as input for the same neuron at the next time step, creating a memory effect. Mathematically, the hidden state $\boldsymbol{h}_t$ at the time step t is computed as:

$$\boldsymbol{h}_t^{(l)} = \sigma[\boldsymbol{w}^{(l)}\boldsymbol{h}_{t-1}^{(l)} + \boldsymbol{w}^{(l-1)}\boldsymbol{h}_t^{(l-1)} + \boldsymbol{b}^{(l)}] \quad (2.11)$$

Therefore, RNNs can handle sequential data. In an RNN, the output at each time step is influenced not only by the current input but also by the previous time steps, creating a temporal relationship within the data. RNN also has memory save and retention through its hidden layer states. Long Short-Term Memory (LSTM) networks and Gated Recurrent Units (GRUs) are advanced variants of RNNs designed to address the issue of vanishing gradients, which can occur in standard RNNs when trying to learn long-term dependencies. This is mainly by means of gating mechanisms that control the flow of information, enabling the network to retain or forget information as needed, which improves the network's ability to learn long-term dependencies. RNNs can remember past inputs and use this memory to influence future predictions. This is essential for tasks like time series forecasting, where future values depend on past observations, or language processing, where the meaning of a word often depends on the previous words in a sentence. All these features give a unique advantage to RNNs, and we will also see the same features in the plasmas' CPNs in Chapter 4.

2.3 General Concept of Training and Physics-Informed Neural Networks

The training of ANN is therefore to find the $\theta = \{\boldsymbol{w}, \boldsymbol{b}\}$ that can always output the correct $\boldsymbol{y}$ according to the input $\boldsymbol{x}$. When the system's complexity exceeds the capability of simple models, a neural network's flexibility allows it to capture intricate patterns in the data, thereby serving as a powerful tool for modeling and optimization. The network effectively learns the underlying relationships in the data, generalizing them to make accurate predictions on unseen inputs. The resulting neural network model can then be used to predict or control the system, even in scenarios where explicit mathematical models are impractical. The most commonly used training algorithm is the backpropagation. This is done through an iterative process that involves calculating gradients of the loss function with respect to the network's parameters and updating those parameters using the prementioned gradient descent technique. For the output layer, the gradients of the loss function with respect to the weights $\boldsymbol{w}^{(L)}$ and the bias $\boldsymbol{b}^{(L)}$ are computed as follows:

$$\frac{\partial \epsilon^{(l)}}{\partial \boldsymbol{w}^{(l)}} = \delta^{(l)}[\boldsymbol{h}^{(l-1)}]^T \tag{2.12a}$$

and

$$\frac{\partial \epsilon^{(l)}}{\partial \boldsymbol{b}^{(l)}} = \delta^{(l)} \tag{2.12b}$$

where

$$\delta^{(l)} = [\boldsymbol{w}^{(l-1)}]^T \delta^{(l+1)} \odot \frac{\partial \sigma[z^{(l)}]}{\partial z^{(l)}} \tag{2.12c}$$

$$z^{(l)} = \boldsymbol{w}^{(l)} \boldsymbol{h}^{(l-1)} + \boldsymbol{b}^{(l)} \tag{2.12d}$$

the superscript T representing the transpose of matrix, and symbol "⊙" is the element-wise product of matrices. Then, the $\boldsymbol{w}$ and $\boldsymbol{b}$ can be updated by

$$\boldsymbol{w}^{(l)} \leftarrow \boldsymbol{w}^{(l)} - \eta \frac{\partial \epsilon^{(l)}}{\partial \boldsymbol{w}^{(l)}} \tag{2.13a}$$

$$\boldsymbol{b}^{(l)} \leftarrow \boldsymbol{b}^{(l)} - \eta \frac{\partial \epsilon^{(l)}}{\partial \boldsymbol{b}^{(l)}} \tag{2.13b}$$

One can see that to achieve the update of a previous layer weights and bias, one must first find the error from the later layer. Therefore, this process must start with the output layer error, and update layer by layer towards the input layer, namely backpropagation training.

However, the control, diagnostics, and analysis on plasma parameters usually do not provide error at the output layer when using ANN. For example, a typical flow chart of analyzing an OES of a plasma is shown in Fig. 2.5. The idea is to train an ANN to predict the chemical composition of a plasma by feeding the ANN with a real-time OES. However, since the complete vector of chemical composition can never be directly detected by means of MS, etc. such as for those cold atmospheric plasma (CAP) jets as introduced in Chapter 1, there is no $\delta^{(L)}$ to begin the backpropagation. The users must use a chemical simulation to solve for a theoretical OES based on the ANN's predictions of the chemical composition and temperature, then, compared the theoretical result with the experimental OES, to acquire the error.[14] This is a typical way of physics-informed neural network (PINN).

PINN relies on using the physical laws and/or any other known physical information to train the ANN.[15,16] If experimental data is used in the training, the method is also called data-driven PINN.[5] The example in Fig. 2.5 is a typical PINN because there is conservation laws used in the chemical simulation module that is to determine the error of the neural network. However, many simpler PINN schemes can have physical laws such as some partial-differential equations (PDEs) directly embedded in the loss function.[5,17,18] In these simpler conditions, backpropagation may be possible. The loss function can be a summation of multiple terms such as

$$\epsilon = \epsilon_{\text{data}} + \lambda \epsilon_{\text{laws}} \tag{2.14}$$

where λ is a critical constant that determines which part is more important to the result. However, in many cases of PINN, it is not possible to embedded laws into the loss function, or even embedded, the derivation is too complicated to be done for backpropagation. In those cases, evolutionary algorithms are recommended.

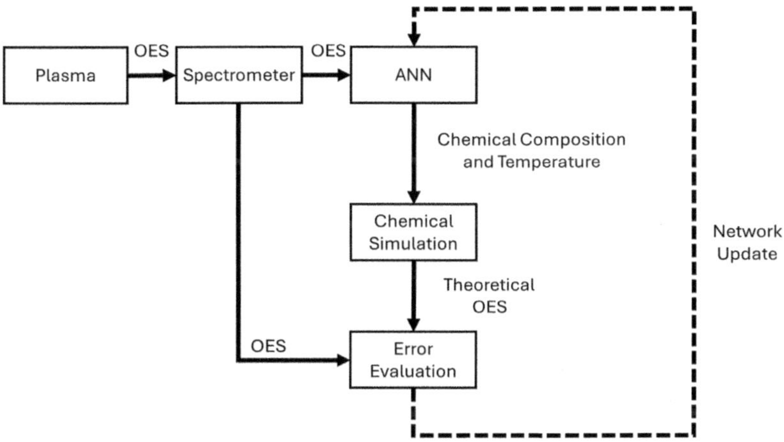

Figure 2.5 A flow-chart example of using ANN to analyze plasma spectrum where no error on the output layer of the neural network is available.

2.4 The Evolutionary Algorithm of Machine Learning Plasmas

Evolutionary algorithms (EAs) are a class of optimization algorithms inspired by the principles of natural selection and evolution.[19-22] They are used to solve complex optimization problems where traditional methods may struggle, particularly in situations involving large search spaces, nonlinearities, or multiple conflicting objectives, as well as when the backpropagation is not possible.[19-22] EAs mimic the process of natural evolution, where a population of candidate solutions evolves over time through operations such as selection, mutation, and crossover.[19-22] The goal is to evolve a population of solutions toward an optimal or near-optimal solution. Generally, in EAs, there is not only one ANN, but many of its mutated versions or offsprings to compete with it. As shown in Fig. 2.6, the total number of individuals involved in the evolutionary training is S_{tot}. No matter what kind of error evaluation is used, better versions can be saved for making the next generation with or without mutations, while others are removed. After a large number of iterations, one will see the ANN evolve. The basic components of an evolutionary algorithm can be summarized as (μ-κ-λ-ρ) evolutionary strategy (ES). The four Greek letters represent the four critical hyperparameters that are used to control the EA. First, μ is the number of individuals making one mating. In EA, the terminology is crossover, or recombination (we will not use "recombination" in this book to avoid confuse with the recombination of ions and electrons). Please note that it does not have to μ = 2 like most of the species mating on Earth or μ = 1 representing the asexual reproduction of single cell organisms or parthenogenesis. In EAs, the gene of each individual is termed θ, which is the set of weight matrices and bias matrices etc. as prementioned. In the process of crossover, each father individual randomly offers a part of the offspring's θ matrices elements. The value of λ represents the total number of offsprings in one iteration. The offspring number generated in each crossover also does not have to be 1. In other words, twins, triplets, etc. are all allowed. There is also a lifetime limit for everyone denoted as κ in the unit of iteration. If an individual is too good to be removed after κ iterations, it will be removed after to prevent premature convergence, in other words, to keep the diversity of the ANNs. Otherwise, that individual will become a stallion and soon, all the versions will become the single stallion's offsprings because crossover between fathers and children is generally allowed. As a result, the EA training will converge to a high-error stage, and the resulting stallion ANN cannot capture essential details of the input data. Such a premature convergence is a major difficulty of EAs but can be solved after carefully adjusting the above hyperparameters. However, please note that premature convergence is a different concept from the under-fitting issue which is a common problem for all ANN training algorithms including EAs, while the premature convergence metaphorically like a biological genetic bottleneck, arising when the diversity of the ANNs is too low.

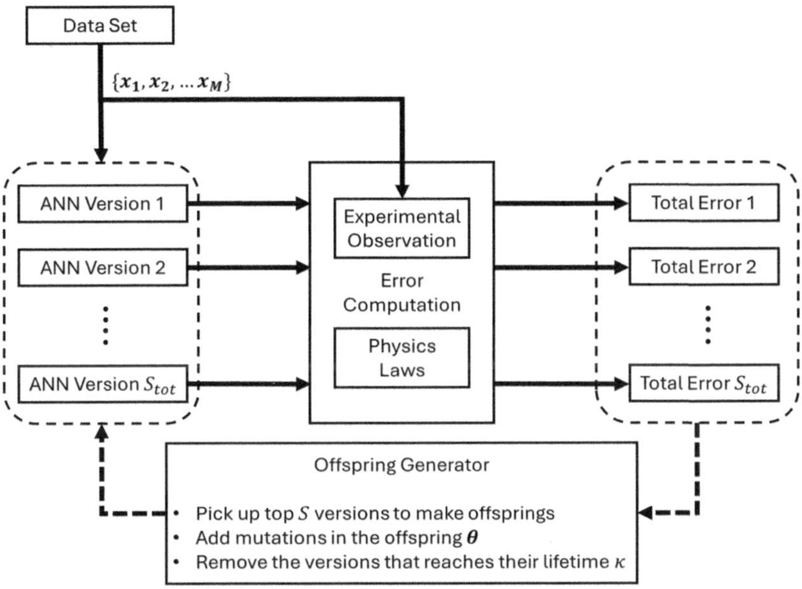

Figure 2.6 The general idea of EAs, where M is the minibatch size, and S_{tot} is the total number of individuals (versions) in the algorithm.

There is another critical concept which is almost applied in all the EAs: mutations. Mutations means adding a small random value on an element in θ. Two hyperparameters are used in mutations: the number of mutations and the mutation range. The former is usually a ratio, it determines how many elements in each matrix of θ are modified. The latter determines the scale of the random value. It can be a random number in a range both above and below the original element, or a Gaussian probability around the original element value with a controlled standard deviation of the Gauss function, etc. An example of using EA to train two ANNs that can diagnose a CAP jet based on its OES and control the jet was published recently. In this example, μ = 1 and the evolution is mainly depending on the mutations. In the next two sections, more engineering examples of using ML to control the plasma systems are introduced.

2.5 Mini Review: Machine Learning Controls in Electric Propulsions

As introduced in Chapter 1, plasma propulsion has emerged as a cutting-edge technology with the potential to revolutionize space exploration. By harnessing the unique properties of ionized gases, plasma propulsion systems produced significantly increased efficiency and speed compared to traditional chemical rockets with greater specific impulse. Space exploration is relentless in its pursuit of innovation, pushing the boundaries of technology to reach the farthest corners of the cosmos. AI encompasses a wide range of techniques which can analyze vast datasets, identify patterns, and make real-time decisions. In recent years, integrating ML techniques into the field of plasma propulsion has opened exciting new avenues for research and development.[23–29] Consequently, ML has emerged as a transformative force, offering a powerful toolkit for optimizing and advancing plasma propulsion systems. By analyzing vast datasets, ML algorithms improve our comprehension of plasma behavior and optimize the performance of these advanced propulsion systems in real time. When applied to plasma propulsion, AI through ML training can address the following key areas: (1) AI can continuously analyze data from onboard sensors and detect anomalies in thruster performance or plasma behavior, such as early detection of issues allows for proactive maintenance and prevents mission failures; (2) ML can adjust thruster parameters in real-time to optimize performance while avoiding instabilities, particularly crucial for long-duration missions where constant human intervention is impractical; (3) AI can compute optimal trajectories, accounting for gravitational forces and other celestial dynamics, to reduce travel time and fuel consumption/waste; (4) ANNs can model complex plasma interactions, aiding in the development of more efficient thrusters and reducing the risk of unforeseen challenges during missions; and (5) AI can optimize propellant usage, extending the operational life of plasma thrusters and reducing the overall cost of missions.

2.5.1 ML Designs and Thrusters Modeling

In recent years, there has been a growing interest in interplanetary missions and the importance of optimizing the design of these thrusters for better performance. Traditionally, thruster design has relied on scaling laws due to the complexity of the underlying physics models. Specifically, in the field of Hall effect thrusters (HETs), a feedforward neural network ensemble (FNN ensemble) is to create a scaling model for HETs.[30] This model is based on an extensive database of HET parameters collected from various research sources.[30] The key advantages of this new model are twofold: firstly, it provides more accurate predictions within the range of input parameters covered by the database compared to existing models; secondly, it does not require specific assumptions about certain parameters, such as back streaming and energy efficiency, and avoids explicit dependencies on discharge power and channel construction.[30] However, it has limitations in terms of extrapolation beyond the database's parameter range, where other models like the SEM (state-of-the-art HET scaling model) may perform better.[30] We can conclude that modern neural networks can accommodate more complex data structures, and additional features can be incorporated into ML models if relevant data becomes available.[30] Lifetime prediction explores methods for estimating the operational lifetime of electric propulsion thrusters.[29] Especially the challenge of measuring wall erosion in on-orbit Hall thrusters, a critical aspect of spacecraft power plants.[29] Since ground-based measurements cannot directly assess wall erosion, the study proposes an inversion method that relies on low-frequency oscillation signals and neural networks.[29] An improved one-dimensional quasi-neutral dynamic fluid mathematical model is developed to create a simulation platform for low-frequency oscillation.[29] By varying the cross-sectional area, the platform generates relevant data that simulates changes in the discharge current low-frequency oscillation waveform under different conditions.[29] A nonlinear neural network model is constructed using the low-frequency oscillation data obtained from the simulation platform.[29] The study analyzes parameters like training function, transfer function, and hidden layer nodes to determine the best model configuration.[29]

Another novel approach, to create a predictive performance model for HETs, is also published recently where less computing power may provide a more flexible HET design technique than a scaling approach.[28] The authors present a new open-source scaling model called HETFit.[28] The key advantages of this approach include the ability to predict thruster performance more accurately within a specified domain, as well as the incorporation of experimental data, which traditional scaling methods do not always consider.[28] To understand the effects and provide a marginally better approximation of ion velocity distribution function (IVDF), Kaganovich et al. mention the multifluid approach in Hall2De.[31] The speed/efficiency of the approach was increased by Jorns's ML methods to obtain a solution for anomalous resistivity.[27] However, currently, such 2D r–z (axisymmetric) solvers with empirical or ML models for electron transport offer the only feasible approach to provide practical support for the design and development of HETs. More recently Jorns explores the application of symbolic regression, a data-driven modeling technique, to understand and predict anomalous electron transport in HETs.[27] The model focuses on the anomalous collision frequency, which plays a crucial role in electron transport and is related to cross-field transport through a generalized Ohm's law.[27] The author used empirical estimates of the anomalous collision frequency derived from three different HETs as training data.[27] A genetic algorithm (GA), a subset of EA, was employed to identify data-driven models expressed as symbolic functions of local plasma properties.[27] These data-driven models not only fit the training data well but also accurately predict anomalous collision frequency values for a test dataset from a fourth thruster not used during training.[27] Comparatively, these data-driven models were evaluated against five existing analytical models for anomalous collision frequency derived from the first principles. It showed that the data-driven models exhibit improved quantitative agreement with both the training and test datasets, highlighting their predictive capability. The study also provided insights into the dominant physical processes governing electron transport in HETs, suggesting that electron Hall drift and ion drift speed may play significant roles, potentially arising from turbulence in the plasma. Despite some limitations, such as approximating kinetic effects with fluid approximation and challenges in extending data-driven models to off-nominal configurations, the author concludes that data-driven modeling holds promise for addressing electron transport in HETs. It suggested that this approach may contribute to the development of predictive Hall thruster models, ultimately enhancing our understanding and control of these propulsion systems. Additionally, Shashkov et al. addressed the challenge of determining the electron anomalous conductivity profile within a Hall thruster when its operational parameters are known through experiments.[26] The approach involved adjusting the parametrically set anomalous conductivity profile until the calculated operational parameters closely matched the experimentally measured data.[26] Please recall that this is exactly what "data-driven" means as we introduced earlier in this chapter. To accomplish this, the researchers utilized a 1D3V hybrid model to compute the operational parameters with parametrically set conductivity.[26] They employed Bayesian optimization with a Gaussian process, which is an ML method capable of handling noisy functions and resolving local minima.[26] The results indicated that for the studied operational modes of KM-88 thruster, there exists a unique solution that corresponds best to the measured operational parameters.[26] The study calculated and compared local plasma parameters for four different operating modes, demonstrating qualitative agreement with the experimental data.[26] The researchers suggested that further improvements in the agreement between calculated and measured parameters could be achieved with a more accurate model of plasma-wall interaction.[26]

However, the fully kinetic self-consistent simulations of plasma thrusters are highly computationally expensive leading to the design process relying majorly on experiments and experiences.[32] Consequently, it had been explored that the application of deep learning techniques, specifically RNN and Wasserstein generative adversarial networks (GAN), in particle-in-cell (PIC) simulations for predicting plasma dynamics in a 1D HET domain.[32] We will have a deeper discussion on PIC in Chapter 3. The primary goal of these authors was to improve computational performance while maintaining high convergence accuracy.[32] The model was trained on a dataset generated using Imperial College's PlasmaSim software, which consisted of 200 spatial nodes and 22.719 simulation time steps.[32] The objective was to create a model capable of predicting plasma parameters across the entire domain over numerous time steps until convergence, effectively capturing the complex dynamics involved in the HET simulation.[32] The study presented results showing the predicted plasma potential distribution based on charge densities as input. Additionally, ion velocity predictions were provided for both single-time step predictions and recursive predictions, where predicted values are iteratively used as input for further predictions. The findings indicated that neural networks could predict plasma parameters with high precision when provided data directly from the PIC model. Error accumulation occurs rapidly in this scenario, leading to a divergence between true and predicted results in a relatively small number of iterations. On the other hand, the study introduces "PlasmaGAN" which is a modified Wasserstein GAN architecture capable of learning the general physical dynamics of ions described by the Boltzmann equation and Maxwell's equations. PlasmaGAN can serve as a deep learning-based simulation accelerator, offering a significant computational speedup of four orders of magnitude compared to traditional PIC simulations.[32] This approach holds promise for accelerating the validation of plasma devices like HETs, overcoming computational scaling limitations associated with classical numerical methods.

In addition to HETs, computational AI models can also simulate and analyze the behavior of pulsed plasma thrusters (PPTs) with growing accuracy. These thrusters are high-specific-impulse electric space propulsion systems. Moreover, the integration of ML into these models has opened new possibilities for improving our comprehension, prediction, and optimization of thruster performance. Firstly, Williams et al.[33] discussed the application of ML to enhance the performance and efficiency of PPTs while reducing computational resource requirements. Two deep ANNs are introduced to predict the shape of current and voltage discharge profiles in parallel plate thrusters based on input parameters like electrode separations and capacitor charge voltage.[33] Both neural network models effectively predict the general trends in current and voltage discharge profiles, but they struggle to capture the seasonality or fine-grained variations in the data (due to the small data set used for training). The more accurate of the two models is referred to as PPTNet.[33] While PPTNet is considered a preliminary model requiring extensive testing and training before implementation, the paper highlights the potential of deep learning techniques in advancing electric propulsion.[33] Secondly, Hossain et al.[34] addressed the need for a reliable model for Ablative Pulsed Plasma Thrusters (APPTs). Their study utilizes an ML approach to develop a model based on experimental APPT data and aims to assess its accuracy and reliability in comparison to existing experimental and simulation models.[34] The research involves conducting 600 test operations under four different operating conditions for APPTs.[34] Experimental and simulation data are obtained, and the comparison results indicate that the newly developed data-driven APPT model exhibits superior accuracy and reliability when compared to the simulated APPT model and real APPTs used in experiments.[34] The ML model achieves an accuracy of 99.99% with a low error order of 10^{-2}, while the simulation model achieves a maximum accuracy of 85% with a higher error order of 10^{2}. Consequently, the data-driven ML model proves to be a successful and highly accurate substitute for modeling APPTs, offering potential applications in verifying APPT system health status and assessing real-time performance.[34] Lastly, Valdes and Khorasani's research paper[24] presented the development of a dynamic neural network-based Fault Detection and Isolation (FDI) scheme for PPTs used in the attitude control subsystem of a satellite. The primary objectives are to detect faults in the multiple thrusters and localize which thruster is malfunctioning.[24] The proposed scheme utilized a multilayer perceptron network with dynamic neurons. It adjusted dynamic network parameters based on input-output data obtained from the electrical circuit of the PPTs to minimize output estimation errors.[24] To evaluate the scheme's effectiveness, a Confusion Matrix approach was employed under various fault scenarios.[24] The study demonstrated that the dynamic network-based FDI scheme was highly reliable in detecting faults within the PPTs, achieving a high level of accuracy and precision, as assessed using the confusion matrix evaluation criterion. In the following section, we will discuss the implementation of ML in gridded ion thrusters.[24]

Gridded Ion Thrusters (GITs) are another type of electric thrusters that play a pivotal role in modern space propulsion systems, offering efficient and reliable means of thrust generation. Traditional methods of optimizing GIT performance involve extensive empirical testing and complex analytical models. However, the integration of ML approaches has introduced a data-driven perspective to enhance our understanding and optimize the design of these advanced propulsion systems. For instance, Sable's research[35] explored the critical role of grid geometry in performance and emphasized its impact on erosion rate and thrust. To optimize these thrusters, designers traditionally employ computationally intensive simulation models of plasma physics.[35] In this study, a baseline simulation model utilized a second-order physics approximation, and the PIC was introduced.[35] Additionally, a neural network was trained using supervised learning based on a subset of previously developed model simulation results.[35] The neural network takes grid geometry and voltages as input and predicts the thrust achieved for each design case.[35] Notably, the neural network demonstrated the ability to accurately predict thrust performance for engine designs it had not encountered during training. The study aimed to assess the effectiveness of neural networks in approximating computational simulations, ultimately improving the efficiency of parametric design studies. Results showed that the neural network effectively predicted thrust performance for a given grid design and significantly reduced simulation time compared to traditional computational models.[35] However, it's essential to note that test cases must fall within the parametric limits of the training design cases. Despite training requirements, neural networks offered a means to enhance the granularity of design space mapping with minimal additional cost. For thrust measurements, there have been researchers focusing on the accurate measurement and prediction of micro-newton-class thrust generated by micro-GITs (μGITs) using a novel torsional pendulum design and high-precision neural network models.[36] Testing of two types of μGITs, radio frequency (RF-μGIT) and direct current (dc-μGIT), revealed their thrust characteristics. The RF-μGIT generated thrust ranging from 0.46 to 2.43 mN, depending on RF power and xenon flow rate, while the dc-μGIT produced thrust from 0.016 to 0.310 mN, influenced by acceleration voltage and flow rate.[36] The study employed ANN and radial basis function neural networks (RBF-NN) to predict the relationship between thrust and input parameters for RF-μGIT.[36] ANN predictions deviated by approximately 10% from measured data, while the RBF-NN model achieved a maximum relative deviation of about 2%.[36] The RBF-NN model was used to create distribution maps of thrust and specific impulse under 600 different working conditions, enhancing the understanding of thruster performance.[36] Additionally, it offered a solution for intelligent regulation of thruster performance based on task requirements.[36] Therefore, we can see that the integration of ML approaches in the study of GITs represents a significant leap forward in the optimization and understanding of these crucial propulsion systems. Also, the research highlighted the potential of ML to accurately predict thrust performance, providing an efficient alternative to computationally intensive simulation models. In the following section, we will discuss the implementation of ML in plasma diagnostics and controls for thrusters.

2.5.2 ML Controls on Thrusters

Another topic is that AI-driven methodologies can improve understanding of plasma behavior and refine control of thruster systems through plasma diagnostics. Like many other ML controls, applying ML models for the diagnostics is the first step in optimizing and advancing propulsion technology. To control, we must first monitor the system's status. The ML-based plasma diagnostics for thruster systems has enabled researchers and engineers to study plasma behavior and enhance the performance of propulsion systems. For instance, Prados[37] explored the application of ANNs in the field of plasma diagnostics. Traditional methods often rely on complex theoretical equations, and this study seeks to evaluate a new approach using ML.[37] Neural network models of the planner Langmuir probe, and the orbital motion limit theory of the probe were created. The research addressed both the forward problem (predicting properties from parameters) and the inverse problem (predicting parameters from properties).[37]

The author successfully implemented ANNs for both forward and inverse problems with remarkable accuracy.[37] This achievement established a theoretical framework for plasma diagnostics using neural networks. It highlighted that ANNs can be applied effectively to various theories, provided that a suitable database is available. While the study successfully predicts key plasma parameters like electron temperature and plasma density with a small margin of error, it suggests that ion temperature has a negligible impact on the I-V curve and can be treated as a constant.[37] In other words, this study demonstrated that ANNs are a powerful tool for plasma diagnostics, offering speed and versatility.

Based on the ML diagnostics, one can start to let ML control the thrusters. Recently, thrust prediction programing using an ANN was described to create an auto-control system for Hall thrusters.[25] As an example, a dataset consists of 73 training data sets, which include operational conditions (such as inner/outer/ trim coil current, xenon mass flow rate, and discharge voltage) and corresponding thrust values.[25] The study investigates how the initial values of weights and parameter updating methods influence the accuracy of thrust prediction and the computational time required.[25] The current ANN model, with 5 layers and 100 nodes, predicts thrust efficiency with a 20% uncertainty due to overfitting.[25] However, with only 73 training data sets, a simpler neural network model with 2 layers and 10 nodes is sufficient to predict thrust within 5% uncertainty, indicating improved performance.[25] Surprisingly, higher expressivity in the neural network led to poorer prediction accuracy due to the limited amount of training data.[25] To address these challenges, the authors adopted a stacked autoencoder (sAE) and optimized the network model using a GA, resulting in a significant reduction in calculation time from 2100 seconds to 100 seconds, without overfitting.[25] The study explored three models: a unit reduction model, an sAE model, and a sAE+GA model.[25] The unit reduction model reduced training time but resulted in a larger mean relative error compared to the previous model.[25] The sAE model improved accuracy compared to the unit reduction model but suffered from overfitting issues.[25] When hyperparameters were optimized using GA, the sAE+GA model demonstrated the shortest training time among the four models, without overfitting.[25] The sAE+GA model was applied to the discharge current with hysteresis, successfully reproducing the measured current, albeit with a slightly larger error than the discharge current without hysteresis.[25] The authors of this work suggest the implementation of various operating conditions for improving their model in the future.[25]

To control miniaturized ion thrusters for space-based gravitational wave detection, we can also consider using a neural network model (NNM). As an example, a recent research aimed to develop a digital twin for miniaturized ion thrusters, enabling real-time monitoring and control of the relationship between grid voltage and extraction current using OES.[38] This digital twin was vital for precise thrust control in gravitational wave detection.[38] The study employed a collisional-radiative model (CRM) to relate plasma parameters within the thruster's discharge chamber to emission spectroscopy.[38] Additionally, an extraction current model (ECM) connected these plasma parameters to the function that describes the relationship between grid voltage and extraction current.[38] The NNM was trained using data generated by these models and was subsequently validated against experimental results from a miniaturized ion thruster.[38] The research demonstrated that the NNM can predict thrust with high accuracy.[38] The difference between the thrust values predicted by the NNM and those obtained experimentally was less than 6%.[38] The study introduces random noise into the training dataset, which proved to enhance the NNM's ability to resist noise in the digital twin model. This noise-resistance feature was crucial for reliable diagnostics in practical applications.

2.5.3 Summary of ML Thrusters

Overall, in this section, by listing some of the recent works of using ML on the space electric propulsion, we can conclude that the integration of ML approaches in electric propulsion systems, with a focus on plasma propulsion, presents a transformative avenue for advancing space exploration capabilities. The applications of ML in diagnostics, control, trajectory optimization, plasma behavior prediction, and propellant management demonstrate its potential to revolutionize the efficiency and performance of advanced propulsion systems. Examining ML for HETs, the studies discussed the versatility of ML, from predicting performance parameters with precision to addressing challenges in lifetime prediction and plasma-wall interactions. The utilization of data-driven models, such as neural networks, offers a powerful toolset for understanding and optimizing the behavior of HETs, contributing to improved design methodologies. For PPTs, ML models emerge as effective tools for predicting discharge profiles, modeling ablation, and implementing fault detection schemes. These applications underline the significance of ML in enhancing the accuracy and reliability of predictive models while enabling real-time performance assessment and system health monitoring. GITs represent a critical component of modern space propulsion, and the integration of ML is instrumental in addressing design challenges. From predicting thrust performance based on grid geometry and voltages to optimizing micro-thrusters, ML demonstrates its efficacy in reducing simulation time and improving predictive accuracy. The studies on miniaturized ion thrusters for space-based gravitational wave detection and plasma diagnostics further emphasize the practical applications of ML in real-time control and accurate predictions. Plasma diagnostics, a crucial aspect of thruster systems, also benefits significantly from ML approaches. The ability of ANNs to predict plasma parameters and create digital twins for precise control showcases the versatility and speed of ML in this domain. Additionally, ML applications in predicting thrust for miniaturized ion thrusters and addressing challenges in plasma-wall interactions highlight its potential to improve reliability and efficiency in practical applications.

2.6 Machine Learning Controls in Plasma Biomedicines

ML has also emerged as a transformative tool across various scientific disciplines, and its integration into biomedical research has opened new frontiers for innovation and efficiency. In particular, the intersection of machine learning with plasma-based biomedical applications signifies a promising avenue for advancing medical treatments. As introduced in Chapter 1, CAPs have gained attention for their diverse therapeutic potential, ranging from cancer treatment to wound healing, sterilization, and antimicrobial applications. Applying an electric field allows for the ionization of gases such as He, Ar, O_2, N_2, and/or air, resulting in the production of electrons, ions, UV, thermal radiation, and various reactive species. However, the complexity of plasma-surface interactions, the variability of plasma sources, and the intricate nature of biological responses present significant challenges in optimizing and standardizing plasma treatments. Using ML algorithms, researchers can look at large datasets, find the most influential parameters, which explain the outcome of treatment and make models for improving plasma treatment technologies. However, it has the potential to improve the treatment results and is one more step for better understanding of plasma-surface or cell interactions which can be further used in biomedicine with a more individual approach. Therefore, integrating machine learning into plasma-based biomedical interventions can reshape treatment protocols, conduct instantaneous diagnoses, and eventually guide the medical care industry towards safe, repeatable, and efficient healthcare practices. Despite the early stages of research in plasma medicine, the integration of AI holds great promise for future applications, offering advancements in treatment optimization, personalized therapies, and a deeper understanding of treatment mechanisms. The combination of plasma medicine with AI applications is highlighted for its potential to enhance treatment effectiveness in various settings, including *in vitro*, *in vivo*, and clinical environments. The review underscores the positive impact of AI on defining "plasma dose"[39–42] and illustrates the significant benefits of integrating plasma and AI in healthcare practices. In this section, we will see that ML offers a unique set of capabilities to decipher intricate patterns, model complex interactions, and enhance our understanding of the dynamic relationships between plasma parameters and biomedical outcomes. We will have a review to look at the current developments in using machine learning techniques in plasma-based biomedical research and applications. We will also introduce in particular how these technologies can be applied to tackle the difficulties facing predictive, diagnostic, and control of CAP-induced effects on cells and living organisms.

2.6.1 Adaptive Plasmas and Plasma Modalities

A specific type of ML-controlled plasma medicine is adaptive plasmas.[14,43,44] The idea is unique because its control and optimization is not only based on the plasma parameter feedback, but also acquiring feedback from the target such as cells and tissues, as schematically concept of adaptive LTP is shown in Fig. 2.7 in which both direct and feedback-based adaptations are illustrated.[14,43,44] The low-temperature plasma treatment *in vivo* or *in vitro* will trigger different cellular responses to plasma and plasma's response to cells via the specific interaction between plasma and cells. The response can be measured by different ways such as Electrical Impedance Spectroscopy (EIS).[14,43,44] The signal based on these responses will be measured, recorded and analyzed by an ML program stored in a microcontroller (MCU). Through such a feedback system, the chemical composition will be modulated through changing the power supply, the feeding gas composition, the distance between plasma source and cells. Additional ways to dynamically control plasma chemistry are power supply modes (such as pulse form, DC, AC, pulse duration from nano-sec to micro-sec, etc.).

In the case of direct feedback or self-adaptation, plasma interaction could lead to self-organization via transition between various discharge modes. In principle, a feedback mechanism on the self-organization can be also used (see Chapter 1 for more details of the self-organization pattern which is a key feature at the plasma-liquid interface). The idea of adaptive plasma for biomedical application, taking cancer therapy as an example, is based on the ability to read the cellular response to plasma in real time and to modify the plasma composition and power via a feedback loop. Specific response of cancer cells to plasma can be utilized as a real time marker as well. This will allow for the modification of the plasma conditions in real time to optimize the plasma effect on cancer cells or normal cells. Due to a distinct effect of cold plasma on normal cells and cancer cells, such a system can lead to an enhanced selectivity of plasma by controlling the generation of reactive species or the potential interaction between plasma and cells.

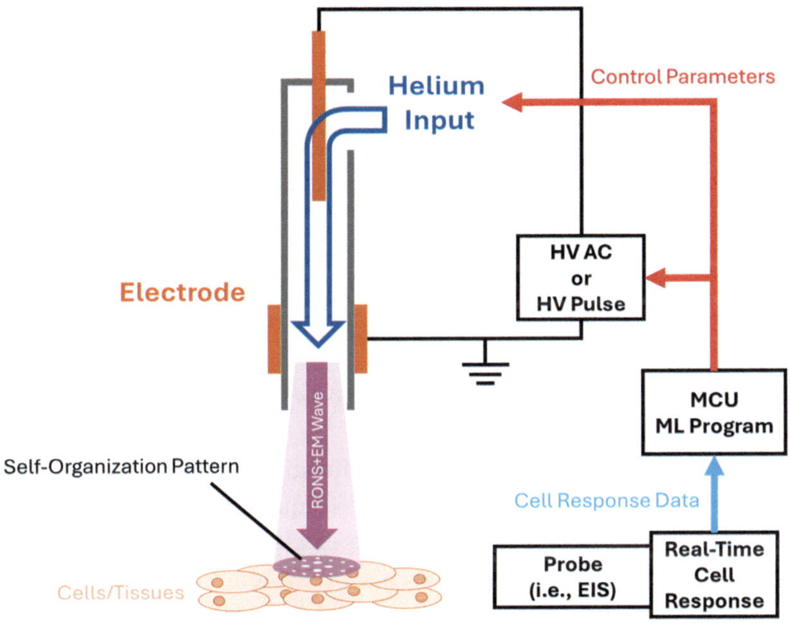

Figure 2.7 Basic concept of adaptive plasma.

In such a type of adaptive plasma, the plasma parameters can even be a black box, and the users can focus on the target status only. However, there is another interesting point in this method. Although the cell response can be measured in real time, the cell viability for example will be affected after 24 hours. Therefore, the ML system is trained to predict the target status in the future based on the real time response signal. Therefore, the adaptive plasma, that is aimed at optimizing the treatment results, requires an algorithm to predict the target status in the future.[45] In other words, a mathematical function between the EIS data and the target status in a specific time delay is needed, which had been studied by Lyu et al.[45] In the next section, we will see the details of the real-time EIS measurement.

2.6.2 Fundamentals of EIS Feedback

To achieve an EIS, one needs to apply an AC voltage on the target with a pair of electrodes, and meanwhile, measure the current in the electrodes (Fig. 2.8). The fundamental concept of EIS is based on the impedance of a material which is defined as

$$Z(\omega)=\frac{U(\omega)}{I(\omega)} \tag{2.15}$$

where Z is the impedance, $\omega = 2\pi f$ is the angular frequency of the voltage U, and I is the current. Impedance is a complex quantity, comprising a real part (resistance), and an imaginary part (reactance) as

$$Z(\omega) = R(\omega) + iX(\omega) \tag{2.16}$$

where R is the resistance, i is the imaginary unit, and X is the reactance. By sweeping a range of frequencies, EIS provides a comprehensive profile of how these electrical properties change, giving insights into the material's or system's behavior. When cells are viable and adhere to the electrodes, or the cells in medium (either suspension or attached at the bottom of a dish), the whole target acts as insulating particles, determining the impedance between the electrodes. Especially, the cells can be considered as a resistor-capacitor system because of its multi-membrane structure.[44,46–48] The equivalent circuit of a cell culture system is usually designed to describe the medium and cells. A typical procedure of EIS analysis is: (1) design an equivalent circuit, (2) measure the spectrum of the sample with a frequency sweep, (3) find the resistance and capacitance values in the equivalent circuit so that the circuit can produce the same spectrum as the sample, (4) make change to the sample such as the plasma treatment, (5) adjust the parameters in the circuit so that the circuit can alter its spectrum in the same way. Finally, in step (5), which circuit parameters must be changed to mimic the spectrum indicates what happened in the real cell system. For example, if one must decrease the resistance in the circuit, which represents the ion density surrounding the cell, so that the circuit can mimic the same EIS spectrum change, we can thus estimate that the ion density near the cell is increased as a result of the resistance decrement.

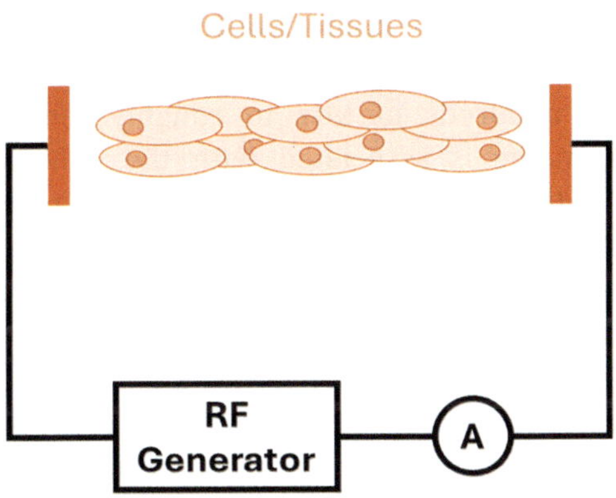

Figure 2.8 A hardware schematic of EIS.

In the EIS spectrum, the frequency-dependent behavior of biological tissues, including cell suspensions, can be characterized by three main dispersion regions: alpha (α) dispersion, beta (β) dispersion, and gamma (γ) dispersion. Each of these dispersions corresponds to different frequency ranges and is associated with specific physiological and structural properties of cells and tissues.

The α dispersion occurs at low frequencies, typically below 10 kHz. This dispersion is primarily associated with the movement of ions in the extracellular medium and the interface between the cell membrane and the surrounding medium. At these low frequencies, the impedance is mainly determined by the resistance of the extracellular medium and the polarization effects at the electrode interface due to the accumulation of charges. In a biomedical system, such as a cell suspension in medium, the α dispersion is influenced by the movement of ions (such as Na^+, K^+, Cl^-) in the extracellular space contributes to the overall impedance. The impedance at low frequencies is also affected by electrode polarization, where ions accumulate near the electrodes, creating a polarization impedance. The α dispersion provides insights into the ionic content and properties of the extracellular medium, including conductivity and permittivity. It is particularly useful for understanding the ionic environment around cells, but it does not provide detailed information about cell membrane properties.

The β dispersion occurs in the frequency range from approximately hundreds of kHz to roughly tens or hundreds of MHz. This range is associated with the polarization of cell membranes. In this region, the cell membrane acts as a capacitor, and its charging and discharging contribute to the impedance spectrum. In a cell suspension, β dispersion is primarily influenced by the cell membrane capacitance and other behaviors related to the membrane. The cell membrane behaves like a dielectric barrier due to its lipid bilayer structure, which separates the intracellular and extracellular ionic environments. The charging and discharging of this membrane capacitance led to characteristic changes in impedance. More importantly, the β dispersion is sensitive to changes in membrane integrity. For instance, when cells are damaged or die, their membranes become more permeable, reducing the capacitance and thus altering the impedance in this frequency range. Recently, Lin et al. measured the human cell membrane oscillation, using laser Michelson interferometer, and the membrane natural frequencies, using an atomic force microscope, at this range, which can verify that the membrane status is related to the β dispersion.[49] Mathematically, the cell membrane can be modeled as a capacitor in parallel with a resistor representing ion channels and other membrane conductance. The impedance of a single cell can thus be described by a simplified model:

$$Z_{\text{cell}} = \left(R_m + \frac{1}{i\omega C_m} \right)^{-1} \tag{2.17}$$

where Z_{cell} is the impedance, R_m is the cell membrane resistance due to the ion channels, and C_m is the membrane (lipid double layer) capacitance. As we can see in biomedical systems, the

β dispersion is especially critical for assessing cell viability and morphology. Changes in the membrane capacitance can indicate cell death, apoptosis, or changes in cell size or shape, making the β dispersion a key indicator in applications such as cytotoxicity assays, drug screening, and tissue engineering.

The γ dispersion occurs at high frequencies, typically from hundreds of MHz to several hundreds of GHz. This dispersion is attributed to the dipolar relaxation of water molecules and other polar molecules within the cells and the extracellular medium. At these high frequencies, the cell membrane becomes less relevant, and the dielectric properties of the cell's internal structures and the medium dominate the impedance response. The γ dispersion usually reflects intracellular and extracellular medium particles. For example, the primary contribution to γ dispersion comes from the relaxation of medium molecules, which have a high dielectric constant and exhibit polarization at these high frequencies. Proteins and other macromolecules in the cytoplasm can also contribute to the γ dispersion due to their polar nature and ability to align with an alternating electric field. For a suspension of cells in a medium, the γ dispersion provides information about the dielectric properties of the intracellular contents and the surrounding medium. For example, the water content, hydration state, and concentrations of macromolecules can be inferred from changes in the impedance spectrum at these frequencies.

The understanding of α, β, and γ dispersions is crucial for interpreting impedance data in biomedical applications. For example, in cell suspensions, these dispersions can provide detailed information on various physiological states. Here, the β dispersion is particularly useful for assessing cell membrane integrity, which is a key marker of cell viability. Damaged or compromised membranes alter the impedance in this frequency range, allowing researchers to detect cell death or damage. Changes in the β dispersion can also indicate alterations in cell size and morphology, which are relevant in studies of cell growth, differentiation, and disease. On the other hand, the γ dispersion is sensitive to the dielectric properties of water and macromolecules inside cells. Variations in intracellular hydration or macromolecule concentrations can be detected by monitoring changes in the γ dispersion.

2.6.3 Mini Review: Machine Learning on Plasma-Based Biomedicine

As we introduced in Chapter 1, high intracellular ROS concentration is a commonly observed hallmark of most cancerous cell types and has been proposed to play a major role in carcinogenesis, as it is intrinsically linked to cell metabolism and growth. Most of the cancers have been observed with increased ROS concentrations which play a major role in carcinogenesis. Cancerous cells over-express antioxidant protein that acts as a defense against the alteration of ROS, allowing for precise control of ROS concentration in the intracellular environment for appropriate functionality of tumors cells. Cellular functions of ROS in cancer depend greatly on the type of radical (e.g., OH radicals, H_2O_2 or HO_2) and generation site (e.g., specific organelles or cytoplasmic versus extracellular). The intracellular ROS signaling should be tightly regulated to restrict ROS-dependent processes involved in tumor growth and at the same time promote ROS-mediated apoptotic signals. Alternatively, it can be argued that therapeutic antioxidants can be used to prevent the initial events of tumor development where ROS are vital. Nevertheless, to selectively target cancer cells, there is a need to better characterize ROS signaling pathways that regulate diverse stress-regulated cellular functions.[50]

We have also introduced in Chapter 1 that CAP is a novel and emerging technology that has shown great promise as a potential therapeutic approach for cancer treatment. CAP is a partially ionized gas composed of various RONS that can selectively target cancer cells while sparing normal tissue. Preclinical studies have demonstrated the anti-tumor effects of CAP in various cancer types, including solid tumors and hematological malignancies.[51] Mechanistic studies have revealed that CAP-induced reactive species lead to the induction of apoptosis, cell cycle arrest, DNA damage, and the disruption of cellular signaling pathways in cancer cells. Additionally, CAP has been shown to sensitize cancer cells to traditional therapies, such as chemotherapy and radiation, potentially enhancing their effectiveness. However, the complete mechanism of action is different in different types of cancer. The complexity of the mechanism of action in various types of cancer presents a considerable challenge due to the inherent diversity in tumor microenvironments. Treatment of cancer is complicated because each tumor displays a different set of attributes and characteristics making it difficult for plasma chemistry to inhibit the tumor growth. Additionally, optimizing the quantity and concentration of species generated through cold atmospheric plasma treatment poses another significant challenge. To address these complex challenges, ML can play an important role by enabling plasma therapy to adapt to the complexity of each tumor by enhancing the adaptability of plasma therapy to individual tumors. ML can also discern the specific requirements for species concentration as well as treatment time tailored to effectively eradicate cancer cells in a dose and time dependent manner. Particularly, ML algorithms can be trained in a time and dose-dependent manner, enabling them to discern the optimal conditions for generating reactive oxygen species that are essential for inducing cell death. This adaptive approach, guided by ML, holds great promise in advancing the precision and efficacy of plasma-based cancer treatments, offering a tailored and nuanced strategy for combating the heterogeneity inherent in cancer biology.

Recent papers have shown the potential of ML to solve the problems associated with the safe and reproducible application of CAPs for medical purposes.[52] The paper outlines three key areas where ML can contribute: data-driven modeling of plasma-surface interactions, real-time diagnostics, and predictive controllers for CAP treatments.[52] Challenges identified include the multitude of reactive agents, poorly understood interactions, and heterogeneous process parameters.[52] It had been suggested that organizing efforts around three sub-aims: quantifying plasma effects, real-time diagnostics, and learning-based control.[52] Parallels are drawn from other medical domains facing similar challenges. We must emphasize the need for standardized clinical data, urging the plasma medicine community to establish widely accepted standards. Data collection procedures should be standardized, and obstacles to ML diffusion must be overcome with user-friendly tools and fail-safe tests. Simplifying data acquisition and control systems is crucial for moving towards point-of-care treatment. Collaboration among plasma scientists, physicians, microbiologists, and data scientists is highlighted as essential for addressing these challenges and realizing the potential of ML in plasma medicine.

To diagnose the plasma treatments, in a way friendly to a real-time ML optimization, Lin et al. employed real-time electro-chemical EIS to monitor and modify plasma treatment.[44] This study was focused on the application of ML and presented an interesting prospect as a tool for discerning the viability of cancer cells and for differentiating between healthy and cancerous cells. As we mentioned in a previous section, this could serve as a valuable feedback mechanism, offering insights into complex processes of cancer inhibition by CAP. By the application of ML, we could gain insight into cellular responses to plasma treatment, specifically for plasma-based cancer therapy. The ability of ML algorithms to process and analyze complex data sets enables the identification of patterns and correlations of cancer cell viability. This proposed use of ML as a feedback mechanism holds promise not only for optimizing the effectiveness of CAP interventions but also for cancer treatment adaptability.[43,53] However, careful dissection of this complex relationship between CAP and cancer cell viability may yield a rich source of information to aid in treatment optimization. Thus, the application of ML in studying the viability of cancer cells under plasma exposure is a promising way to further develop knowledge on cancer and plasma medicine. Such an approach may reveal new perspectives about the best strategies for general cancer treatment and the development of tumor type or patient specific treatment regimens.[54]

Conventional sterilization methods for medical devices or surfaces typically involve heat, UV and gamma ray irradiation, and chemical gases,[55] Please recall that these are all the outputs as shown in Fig. 2.1 at the beginning of this chapter. While generally effective, these methods have drawbacks such as efficacy, alteration of material properties, ionizing radiation or toxic fume exposure of personnel, and the requirement of high temperature and high-pressure environments. CAP technology, operating under atmospheric conditions below 40°C, offers a promising alternative.[55] CAP permits the disinfection or sterilization of thermosensitive materials and has shown efficacy in *in vitro, ex vivo*, and *in vivo* applications.[55] CAP's ability to generate RONS also makes it a valuable tool for surface decontamination and hand disinfection in public health and hospital care. Hospital-acquired infections (HAIs) remain a significant concern, and incident rates continue to grow. Sterilization and material treatment by CAP presents a potential solution, addressing the decline in pathogen susceptibility and associated economic costs. The rise in antimicrobial resistance among pathogens, including methicillin-resistant *Staphylococcus aureus* (MRSA), vancomycin-resistant *Enterococci* (VRE), carbapenemase-producing *Enterobacterales* (CPE), and extended-spectrum beta-lactamase-producing Gram-negative bacteria like *Escherichia coli*, poses a significant concern.[56] HAIs have substantial economic implications, leading to increased morbidity, mortality, illness, and direct costs estimated to rise. Consequently, there is an urgent need for new strategies in the prevention and treatment of HAIs. Recent research has demonstrated the efficacy of CAP technology in deactivating a variety of microbial pathogens, encompassing bacteria, fungi, and viruses, while also exhibiting the capability to degrade toxins.[55] This versatile technology proves effective in disinfecting medical and dental devices, as well as agricultural products. Practical applications range from sterilizing therapeutic medical instruments to enhancing crop yields and preserving food.[57] Emphasizing the potential applications of plasma technology for microbial inactivation and toxin degradation, we can underscore its broad applicability across the resistance hierarchy of microorganisms. Despite comprehensive studies on various microorganism categories, including those highly resistant and susceptible, investigations on the CAP treatment of viral particles are yet to be reported. However, the usage of CAP technology extends beyond agriculture and medicine, encompassing environmental applications such as water purification, remediation, and exhaust gas treatment. The precise mechanism underlying the sterilization process facilitated by CAP remains elusive, yet prevailing literature emphasizes the significant role of RONS as primary contributing factors. ROS species, particularly singlet oxygen and hydrogen peroxide, have been identified as potent agents for bacterial cell damage by oxidation of cytoplasmic membrane, lipid peroxidation, and subsequent oxidative damage to DNA strands of *E. coli*.[58,59] The presence of RNS is found to enhance inactivation efficacy when combined with ROS, emphasizing the importance of oxygen blend in working gases.[60,61] Reactive species generated in CAP, including charged particles, electrons, and ions, contribute to bacterial inactivation through mechanisms such as cell membrane damage, erosion, DNA damage, rearrangement of chemical bonds, formation of lesions, and penetration of plasma toxic compounds into bacterial cells.[59,62] Charged particles, classified as direct and indirect treatments, play a crucial role in mechanical disruption of bacterial cell membranes.[63] Indirect treatments utilize distance or metal mesh to prevent direct contact, while direct contact involves charged particles accumulating on the surface, causing electrostatic stress and cell membrane perforation via electroporation, and potentially the bioelectric effect.[64]

Therefore, the complexity of CAP and CAP treated media require the ML's involvement to diagnose, analysis, and optimize the plasma's output. For example, Ozdemir et al. have studied sterilization using plasma-activated liquids (PALs),[56] a.k.a. plasma-activated medium (PAM) in many other works, or specifically, there are also plasma-activated water (PAW) introduced. This research explores the application of ML in predicting the antimicrobial activity of plasma-activated liquids (PALs), which have proven broad-spectrum antimicrobial effects. The complexity of PALs' antimicrobial effects, influenced by various parameters, makes direct comparisons challenging.[56] However, the authors mentioned that some of the key issues for sterilization are differences in the plasma devices and generation systems, electrical parameters, electrode geometry and type, liquid composition, and treatment volume, among others, make it challenging to compare the antimicrobial efficacies of PALs across different research laboratories. Additionally, there is a lack of a standardized method for assessing, comparing, and predicting the antimicrobial effectiveness of PALs generated by diverse plasma systems. Furthermore, the inhibitory activity of the generated PALs may vary based on the specific liquids and strains of microorganisms involved.[56] Consequently, there is a pressing need for a method that can predict the antimicrobial potency and efficacy of PALs, ensuring the achievement of desired antimicrobial activity.[56] The study employs comparative supervised ML models, utilizing data from 33 relevant articles and extracting 12 features related to PAL-microorganism interactions. After normalization, encoding, and resampling, classification and regression ML methods are applied, involving 16 classifiers and 14 regressors. Two robust cross-validation strategies are employed, revealing that hyperparameter-optimized Random Forest Classifier (oRFC) and Random Forest Regressor (oRFR) outperform other models, achieving a test accuracy of 82.68% for classification and an R2 score of 0.75 for regression. The identified key features align with existing literature on PALs. The study represents the first attempt to use ML to predict PAL antimicrobial efficacy, providing valuable insights into plasma treatment parameters. The ML models demonstrate high predictive accuracy and may aid in defining a "plasma dose" for various CAP applications in medicine and biology, despite certain limitations and the need for further experimental validation.

2.7 Concluding Remarks

In this chapter, we introduced the interactions among plasma parameters in a plasma system based on plasma physics and the human interface such as the controls and plasma diagnostics. Next, the fundamental idea of control and optimization is introduced with an example of how the error can be decreased by means of error gradient descent method. However, since the mathematical model of a complicated plasma system may not be available, we can choose to use ANN as a high-order polynomial model which has a high fitting capability to map between the input and output space. After the introduction of the math of backpropagation, we introduce EA. This becomes effective when the error on the output layer of an ANN is not available.

We also made a mini review on ML with plasma-based space propulsion and biomedicine are given in the final two sections in this chapter after introducing the theories of controls and ML. However, there is still a major unknown that is considered as a black box in this chapter: the collisions in Fig. 2.1. Therefore, in the next chapter, we will go into the microscopic world, start with the collision theory, and finally discuss the CPN of plasmas to see a newly discovered and revolutionary feature of it.

References

1. Pei, X.; Wu, S.; Xian, Y.; Lu, X.; Pan, Y. On OH Density of an Atmospheric Pressure Plasma Jet by Laser-Induced Fluorescence. *IEEE Trans. Plasma Sci.* **2014**, *42* (5), 1206–1210. https://doi.org/10.1109/TPS.2014.2306921.
2. Settersten, T. B.; Dreizler, A.; Farrow, R. L. Temperature- and Species-Dependent Quenching of CO B Probed by Two-Photon Laser-Induced Fluorescence Using a Picosecond Laser. *J. Chem. Phys.* **2002**, *117* (7), 3173–3179. https://doi.org/10.1063/1.1493198.
3. Amorim, J.; Baravian, G.; Jolly, J. Laser-Induced Resonance Fluorescence as a Diagnostic Technique in Non-Thermal Equilibrium Plasmas. *J. Phys. Appl. Phys.* **2000**, *33* (9), R51–R65. https://doi.org/10.1088/0022-3727/33/9/201.
4. Voigt, S.; Orphal, J.; Burrows, J. P. The Temperature and Pressure Dependence of the Absorption Cross-Sections of NO_2 in the 250–800 Nm Region Measured by Fourier-Transform Spectroscopy. *J. Photochem. Photobiol. Chem.* **2002**, *149* (1–3), 1–7. https://doi.org/10.1016/S1010-6030(01)00650-5.
5. Lin, L.; Gershman, S.; Raitses, Y.; Keidar, M. Data-Driven Prediction of the Output Composition of an Atmospheric Pressure Plasma Jet. *J. Phys. Appl. Phys.* **2024**, *57* (1), 015203. https://doi.org/10.1088/1361-6463/acfcc7.
6. Belostotskiy, S. G.; Khandelwal, R.; Wang, Q.; Donnelly, V. M.; Economou, D. J.; Sadeghi, N. Measurement of Electron Temperature and Density in an Argon Microdischarge by Laser Thomson Scattering. *Appl. Phys. Lett.* **2008**, *92* (22), 221507. https://doi.org/10.1063/1.2939437.
7. Dzierżęga, K.; Mendys, A.; Pokrzywka, B.; Zawadzki, W.; Pellerin, S. Simultaneous Measurement of Electron and Heavy Particle Temperatures in He Laser-Induced Plasma by Thomson and Rayleigh Scattering. *Appl. Phys. Lett.* **2013**, *102* (13), 134108. https://doi.org/10.1063/1.4801467.
8. Muraoka, K.; Kono, A. Laser Thomson Scattering for Low-Temperature Plasmas. *J. Phys. Appl. Phys.* **2011**, *44* (4), 043001. https://doi.org/10.1088/0022-3727/44/4/043001.
9. Shneider, M. N.; Miles, R. B. Microwave Diagnostics of Small Plasma Objects. *J. Appl. Phys.* **2005**, *98* (3), 033301. https://doi.org/10.1063/1.1996835.

10. Lin, L.; Lyu, Y.; Shneider, M. N.; Keidar, M. Average Electron Temperature Estimation of Streamer Discharge in Ambient Air. *Rev. Sci. Instrum.* **2018**, *89* (11), 113502. https://doi.org/10.1063/1.5027836.

11. Lin, L.; Lyu, Y.; Trink, B.; Canady, J.; Keidar, M. Cold Atmospheric Helium Plasma Jet in Humid Air Environment. *J. Appl. Phys.* **2019**, *125* (15), 153301. https://doi.org/10.1063/1.5086177.

12. Barabás, I.; Todoruţ, A.; Cordoş, N.; Molea, A. Current Challenges in Autonomous Driving. *IOP Conf. Ser. Mater. Sci. Eng.* **2017**, *252*, 012096. https://doi.org/10.1088/1757-899X/252/1/012096.

13. Levinson, J.; Askeland, J.; Becker, J.; Dolson, J.; Held, D.; Kammel, S.; Kolter, J. Z.; Langer, D.; Pink, O.; Pratt, V.; Sokolsky, M.; Stanek, G.; Stavens, D.; Teichman, A.; Werling, M.; Thrun, S. Towards Fully Autonomous Driving: Systems and Algorithms. In *2011 IEEE Intelligent Vehicles Symposium (IV)*; IEEE: Baden-Baden, Germany, 2011; pp. 163–168. https://doi.org/10.1109/IVS.2011.5940562.

14. Lin, L.; Yan, D.; Lee, T.; Keidar, M. Self-Adaptive Plasma Chemistry and Intelligent Plasma Medicine. *Adv. Intell. Syst.* **2022**, *4* (3), 2100112. https://doi.org/10.1002/aisy.202100112.

15. Cai, S.; Mao, Z.; Wang, Z.; Yin, M.; Karniadakis, G. E. Physics-Informed Neural Networks (PINNs) for Fluid Mechanics: A Review. *Acta Mech. Sin.* **2021**, *37* (12), 1727–1738. https://doi.org/10.1007/s10409-021-01148-1.

16. Lawal, Z. K.; Yassin, H.; Lai, D. T. C.; Che Idris, A. Physics-Informed Neural Network (PINN) Evolution and Beyond: A Systematic Literature Review and Bibliometric Analysis. *Big Data Cogn. Comput.* **2022**, *6* (4), 140. https://doi.org/10.3390/bdcc6040140.

17. Meng, X.; Li, Z.; Zhang, D.; Karniadakis, G. E. PPINN: Parareal Physics-Informed Neural Network for Time-Dependent PDEs. *Comput. Methods Appl. Mech. Eng.* **2020**, *370*, 113250. https://doi.org/10.1016/j.cma.2020.113250.

18. Karniadakis, G. E.; Kevrekidis, I. G.; Lu, L.; Perdikaris, P.; Wang, S.; Yang, L. Physics-Informed Machine Learning. *Nat. Rev. Phys.* **2021**, *3* (6), 422–440.

19. Bäck, T.; Schwefel, H.-P. An Overview of Evolutionary Algorithms for Parameter Optimization. *Evol. Comput.* **1993**, *1* (1), 1–23. https://doi.org/10.1162/evco.1993.1.1.1.

20. Deb, K.; Anand, A.; Joshi, D. A Computationally Efficient Evolutionary Algorithm for Real-Parameter Optimization. *Evol. Comput.* **2002**, *10* (4), 371–395. https://doi.org/10.1162/106365602760972767.

21. Eiben, A. E.; Hinterding, R.; Michalewicz, Z. Parameter Control in Evolutionary Algorithms. *IEEE Trans. Evol. Comput.* **1999**, *3* (2), 124–141. https://doi.org/10.1109/4235.771166.

22. Michalewicz, Z.; Schoenauer, M. Evolutionary Algorithms for Constrained Parameter Optimization Problems. *Evol. Comput.* **1996**, *4* (1), 1–32. https://doi.org/10.1162/evco.1996.4.1.1.

23. Taploo, A.; Soni, V.; Solomon, H.; McCraw, M.; Lin, L.; Spinelli, J.; Shepard, S. P.; Solares, S. D.; Keidar, M. Characterization of a Circular Arc Electron Source for a Self-Neutralizing Air-Breathing Plasma Thruster. *J. Electr. Propuls.* **2023**, *2* (1), 21. https://doi.org/10.1007/s44205-023-00058-7.

24. Valdes, A.; Khorasani, K. Dynamic Neural Network-Based Pulsed Plasma Thruster (PPT) Fault Detection and Isolation for the Attitude Control System of a Satellite. In *2008 IEEE International Joint Conference on Neural Networks (IEEE World Congress on Computational Intelligence)*; IEEE: Hong Kong, China, 2008; pp. 2689–2695. https://doi.org/10.1109/IJCNN.2008.4634175.

25. Kawazu, M.; Yamamoto, N.; Chono, M.; Fuchigami, H.; Morita, T. Prediction of Thruster Performance in Hall Thrusters Using Neural Network with Auto Encoder. *Trans. Jpn. Soc. Aeronaut. SPACE Sci. Aerosp. Technol. Jpn.* **2021**, *19* (5), 760–765. https://doi.org/10.2322/tastj.19.760.

26. Shashkov, A.; Tyushev, M.; Lovtsov, A.; Tomilin, D.; Kravchenko, D. Machine Learning-Based Method to Adjust Electron Anomalous Conductivity Profile to Experimentally Measured Operating Parameters of Hall Thruster. *Plasma Sci. Technol.* **2022**, *24* (6), 065502. https://doi.org/10.1088/2058-6272/ac59e1.

27. Jorns, B. Predictive, Data-Driven Model for the Anomalous Electron Collision Frequency in a Hall Effect Thruster. *Plasma Sources Sci. Technol.* **2018**, *27* (10), 104007. https://doi.org/10.1088/1361-6595/aae472.

28. Korolev, K. Hall Effect Thruster Design via Deep Neural Network for Additive Manufacturing. arXiv March 14, 2023. http://arxiv.org/abs/2303.08227 (accessed 2024-08-31).

29. Han, K.; Xie, F.; Wang, Y.; Zhang, L.; Yu, M.; Wang, J.; Wang, Y.; Wan, J. Study on Inversion Method of Wall Erosion Information of On-Orbit Hall Thruster Based on Low-Frequency Oscillation Signals and Neural Networks. *Heliyon* **2022**, *8* (11), e11616. https://doi.org/10.1016/j.heliyon.2022.e11616.

30. Plyashkov, Y. V.; Shagayda, A. A.; Kravchenko, D. A.; Lovtsov, A. S.; Ratnikov, F. D. On Scaling of Hall-Effect Thrusters Using Neural Nets. *J. Propuls. Power* **2022**, *38* (6), 935–944. https://doi.org/10.2514/1.B38592.

31. Kaganovich, I. D.; Smolyakov, A.; Raitses, Y.; Ahedo, E.; Mikellides, I. G.; Jorns, B.; Taccogna, F.; Gueroult, R.; Tsikata, S.; Bourdon, A.; Boeuf, J.-P.; Keidar, M.; Powis, A. T.; Merino, M.; Cappelli, M.; Hara, K.; Carlsson, J. A.; Fisch, N. J.; Chabert, P.; Schweigert, I.; Lafleur, T.; Matyash, K.; Khrabrov, A. V.; Boswell, R. W.; Fruchtman, A. Physics of E × B Discharges Relevant to Plasma Propulsion and Similar Technologies. *Phys. Plasmas* **2020**, *27* (12), 120601. https://doi.org/10.1063/5.0010135.

32. Mörtl, M.; Knoll, A.; Williams, V.; Shaw, P.; Argyriou, V.; Zamattio, J.; Pugliese, L. Enhancing Hall Effect Thruster Simulations with Deep Recurrent Networks. In *36th International Electric Propulsion Conference, (Vienna, Austria)*, 2019.

33. Williams, V.; Argyriou, V.; Shaw, P.; Montag, C.; Herdrich, G.; Knoll, A.; Mörtl, M. Development of PPTNet a Neural Network for the Rapid Prototyping of Pulsed Plasma Thrusters. In *Proc. 36th Int. Electr. Propuls. Conf.*; 2019; pp. 1–17.

34. Hossain, N.; Wang, N.; Sun, G.; Li, H.; Wu, Z. A Reliable Data-Driven Model for Ablative Pulsed Plasma Thruster. *Aerosp. Sci. Technol.* **2020**, *105*, 105953.

35. Sable, S. D. Design of Ion Thruster Optics Simulator and a Neural Network Emulator. **2022**.

36. Wang, H.; Wang, W.; Yan, J.; Fu, C.; Liu, W. Measurement and Prediction of Micronewton Class Thrust of Electric Propulsion Based on the Torsional Pendulum and Machine Learning Technique. *IEEE Trans. Instrum. Meas.* **2023**, *72*, 1–14. https://doi.org/10.1109/TIM.2022.3225035.

37. Villegas Prados, D. Evaluation of Artificial Neural Network Method Applied to Probe Diagnostics in Plasma. **2018**.

38. Jia, J.-W.; Zou, R.-Q.; Luo, Q.; Zhu, X.-M.; Wang, Y.-F.; Wu, Y.-J.; Lang, H.; Zhang, W.-J.; Dong, X.-J. A Neural Network Model Relating Extraction Current Characteristics with Optical Emission Spectra for Digital Twin of Miniaturized Ion Thrusters. In *2023 IEEE 10th International Workshop on Metrology for AeroSpace (MetroAeroSpace)*; IEEE: Milan, Italy, 2023; pp 666–671. https://doi.org/10.1109/MetroAeroSpace57412.2023.10190002.

39. Bonzanini, A. D.; Paulson, J. A.; Graves, D. B.; Mesbah, A. Safe Dose Delivery in Fast Sampling Atmospheric Plasmas Using Projection-Based Approximate Economic MPC. *Proc. IFAC World Congr.* 2020; pp 5353–5359.
40. Cheng, H.; Xu, J.; Li, X.; Liu, D.; Lu, X. On the Dose of Plasma Medicine: Equivalent Total Oxidation Potential (ETOP). *Phys. Plasmas* **2020**, *27* (6), 063514. https://doi.org/10.1063/5.0008881.
41. Gidon, D.; Graves, D. B.; Mesbah, A. Effective Dose Delivery in Atmospheric Pressure Plasma Jets for Plasma Medicine: A Model Predictive Control Approach. *Plasma Sources Sci. Technol.* **2017**, *26* (8), 085005. https://doi.org/10.1088/1361-6595/aa7c5d.
42. Gidon, D.; Graves, D. B.; Mesbah, A. Predictive Control of 2D Spatial Thermal Dose Delivery in Atmospheric Pressure Plasma Jets. *Plasma Sources Sci. Technol.* **2019**, *28* (8), 085001. https://doi.org/10.1088/1361-6595/ab2c66.
43. Keidar, M. Adaptive Low-Temperature Plasmas. *Plasma Phys. Control. Fusion* **2024**, *66* (3), 035010. https://doi.org/10.1088/1361-6587/ad15f1.
44. Lin, L.; Hou, Z.; Yao, X.; Liu, Y.; Sirigiri, J. R.; Lee, T.; Keidar, M. Introducing Adaptive Cold Atmospheric Plasma: The Perspective of Adaptive Cold Plasma Cancer Treatments Based on Real-Time Electrochemical Impedance Spectroscopy. *Phys. Plasmas* **2020**, *27* (6), 063501. https://doi.org/10.1063/5.0003528.
45. Lyu, Y.; Lin, L.; Gjika, E.; Lee, T.; Keidar, M. Mathematical Modeling and Control for Cancer Treatment with Cold Atmospheric Plasma Jet. *J. Phys. Appl. Phys.* **2019**, *52* (18), 185202. https://doi.org/10.1088/1361-6463/ab061d.
46. Bounik, R.; Cardes, F.; Ulusan, H.; Modena, M. M.; Hierlemann, A. Impedance Imaging of Cells and Tissues: Design and Applications. *BME Front.* **2022**, *2022*, 9857485. https://doi.org/10.34133/2022/9857485.
47. Grossi, M.; Riccò, B. Electrical Impedance Spectroscopy (EIS) for Biological Analysis and Food Characterization: A Review. *J. Sens. Sens. Syst.* **2017**, *6* (2), 303–325. https://doi.org/10.5194/jsss-6-303-2017.
48. K'Owino, I. O.; Sadik, O. A. Impedance Spectroscopy: A Powerful Tool for Rapid Biomolecular Screening and Cell Culture Monitoring. *Electroanalysis* **2005**, *17* (23), 2101–2113. https://doi.org/10.1002/elan.200503371.

49. Lin, L.; McCraw, M. R.; Uluutku, B.; Liu, Y.; Yan, D.; Soni, V.; Horkowitz, A.; Yao, X.; Limanowski, R.; Solares, S. D.; Beilis, I. I.; Keidar, M. Cell Membrane Oscillations under Radiofrequency Electromagnetic Modulation. *Langmuir* **2023**, *39* (9), 3320–3331. https://doi.org/10.1021/acs.langmuir.2c03181.

50. Liou, G.-Y.; Storz, P. Reactive Oxygen Species in Cancer. *Free Radic. Res.* **2010**, *44* (5), 479–496. https://doi.org/10.3109/10715761003667554.

51. Soni, V.; Adhikari, M.; Simonyan, H.; Lin, L.; Sherman, J. H.; Young, C. N.; Keidar, M. In Vitro and In Vivo Enhancement of Temozolomide Effect in Human Glioblastoma by Non-Invasive Application of Cold Atmospheric Plasma. *Cancers* **2021**, *13* (17), 4485. https://doi.org/10.3390/cancers13174485.

52. Bonzanini, A. D.; Shao, K.; Graves, D. B.; Hamaguchi, S.; Mesbah, A. Foundations of Machine Learning for Low-Temperature Plasmas: Methods and Case Studies. *Plasma Sources Sci. Technol.* **2023**, *32* (2), 024003. https://doi.org/10.1088/1361-6595/acb28c.

53. Keidar, M.; Yan, D.; Beilis, I. I.; Trink, B.; Sherman, J. H. Plasmas for Treating Cancer: Opportunities for Adaptive and Self-Adaptive Approaches. *Trends Biotechnol.* **2018**, *36* (6), 586–593. https://doi.org/10.1016/j.tibtech.2017.06.013.

54. Lin, L.; Keidar, M. A Map of Control for Cold Atmospheric Plasma Jets: From Physical Mechanisms to Optimizations. *Appl. Phys. Rev.* **2021**, *8* (1), 011306. https://doi.org/10.1063/5.0022534.

55. Laroussi, M. Low Temperature Plasma-Based Sterilization: Overview and State-of-the-Art. *Plasma Process. Polym.* **2005**, *2* (5), 391–400. https://doi.org/10.1002/ppap.200400078.

56. Özdemir, M. A.; Özdemir, G. D.; Gül, M.; Güren, O.; Ercan, U. K. Machine Learning to Predict the Antimicrobial Activity of Cold Atmospheric Plasma-Activated Liquids. *Mach. Learn. Sci. Technol.* **2023**, *4* (1), 015030. https://doi.org/10.1088/2632-2153/acc1c0.

57. Birania, S.; Attkan, A. K.; Kumar, S.; Kumar, N.; Singh, V. K. Cold Plasma in Food Processing and Preservation: A Review. *J. Food Process Eng.* **2022**, *45* (9), e14110. https://doi.org/10.1111/jfpe.14110.

58. Joshi, S. G.; Cooper, M.; Yost, A.; Paff, M.; Ercan, U. K.; Fridman, G.; Friedman, G.; Fridman, A.; Brooks, A. D. Nonthermal Dielectric-Barrier Discharge Plasma-Induced Inactivation Involves Oxidative DNA Damage and Membrane Lipid Peroxidation in *Escherichia Coli*. *Antimicrob. Agents Chemother.* **2011**, *55* (3), 1053–1062. https://doi.org/10.1128/AAC.01002-10.

59. Gallagher, M. J.; Vaze, N.; Gangoli, S.; Vasilets, V. N.; Gutsol, A. F.; Milovanova, T. N.; Anandan, S.; Murasko, D. M.; Fridman, A. A. Rapid Inactivation of Airborne Bacteria Using Atmospheric Pressure Dielectric Barrier Grating Discharge. *IEEE Trans. Plasma Sci.* **2007**, *35* (5), 1501–1510. https://doi.org/10.1109/TPS.2007.905209.

60. Boxhammer, V.; Morfill, G. E.; Jokipii, J. R.; Shimizu, T.; Klämpfl, T.; Li, Y.-F.; Köritzer, J.; Schlegel, J.; Zimmermann, J. L. Bactericidal Action of Cold Atmospheric Plasma in Solution. *New J. Phys.* **2012**, *14* (11), 113042. https://doi.org/10.1088/1367-2630/14/11/113042.

61. Sureshkumar, A.; Sankar, R.; Mandal, M.; Neogi, S. Effective Bacterial Inactivation Using Low Temperature Radio Frequency Plasma. *Int. J. Pharm.* **2010**, *396* (1–2), 17–22.

62. Moreau, M.; Orange, N.; Feuilloley, M. G. J. Non-Thermal Plasma Technologies: New Tools for Bio-Decontamination. *Biotechnol. Adv.* **2008**, *26* (6), 610–617.

63. Dobrynin, D.; Fridman, G.; Friedman, G.; Fridman, A. Physical and Biological Mechanisms of Direct Plasma Interaction with Living Tissue. *New J. Phys.* **2009**, *11* (11), 115020. https://doi.org/10.1088/1367-2630/11/11/115020.

64. Laroussi, M. Low-Temperature Plasmas for Medicine? *IEEE Trans. Plasma Sci.* **2009**, *37* (6), 714–725. https://doi.org/10.1109/TPS.2009.2017267.

Chapter 3

Collisions in Plasmas

Li Lin and Zichao Hou

School of Engineering and Applied Science, The George Washington University, Washington, DC 20052, USA

lilin@email.gwu.edu

3.1 Collisions in Classic Physics Regime

Unlike the particles in solids, liquids, or gases, the charged particles in plasmas interact with one another through long-range electromagnetic forces, leading to a rich array of collision dynamics. In plasma, collisions occur between different species of particles, such as electron-electron, electron-ion, ion-ion, and neutral particle interactions. These so-called Coulomb collisions play a critical role in determining the transport properties of the plasma, such as electrical conductivity, thermal conductivity, and viscosity. The frequency and outcome of these collisions depend on factors like particle energy, density, and temperature, which can vary widely across different types of plasmas, from the low-temperature plasmas used in industrial applications to the high-temperature plasmas found in astrophysical environments.[1–4] As mentioned in the previous chapters, inelastic collisions are the microscopic mechanism of chemistry. For example, in the

Machine Learning Plasmas and the Neuromorphic Plasma Chemistry
Edited by Li Lin and Michael Keidar

ISBN 978-981-5129-84-7 (Hardcover), 978-1-003-71013-4 (eBook)
www.jennystanford.com

DOI: 10.1201/9781003710134-3

previous chapters, we have introduced that in low-temperature plasmas, typically used in applications like plasma medicine, plasma etching or plasma-enhanced chemical vapor deposition (PECVD), collisions between electrons and neutral atoms or molecules are particularly important.[5-9] These collisions can lead to ionization, excitation, and dissociation processes, which are essential for sustaining the plasma and driving chemical reactions. In contrast, in high-temperature plasmas, such as those found in fusion reactors or stellar interiors, collisions among ions are more prevalent, leading to nuclear reactions and energy transfer processes that sustain the extreme conditions within these environments.[10,11] Therefore, understanding particle collisions in plasmas is crucial for controlling and optimizing plasma behavior in both natural and engineered systems. Whether in the applications of energy generation, space propulsion, or material processing, the study of these collisions provides insight into the complex interactions that govern plasma dynamics. Let's start with the collision theory of particles.

The Coulomb force between two charged particles is given by:

$$\boldsymbol{F} = \frac{q_1 q_2}{4\pi\varepsilon_0} \frac{\boldsymbol{r}}{r^3} \tag{3.1}$$

where $\boldsymbol{F}$ is the force, q_1 and q_2 are the two charges, ε_0 is the permittivity of the space, and $\boldsymbol{r}$ is the vector from one particle to the other. To analyze collisions, we introduce the concept of the impact parameter b, which is the perpendicular distance between the initial trajectory of a particle and the center of the force field (the other particle). The scattering angle θ describes the deflection of the particle due to the collision. The collision of neutral particles is illustrated in Fig. 3.1.

In the view of classic physics, one can consider the particles as solid spheres, or in other words hard spheres. Therefore, the critical impact parameter $b_c = r_1 + r_2$ which is the lower limit to ensure two solid spheres can directly contact for the collision. The collision cross section can thus be defined as[4]

$$\sigma = \pi b_c^2 \tag{3.2}$$

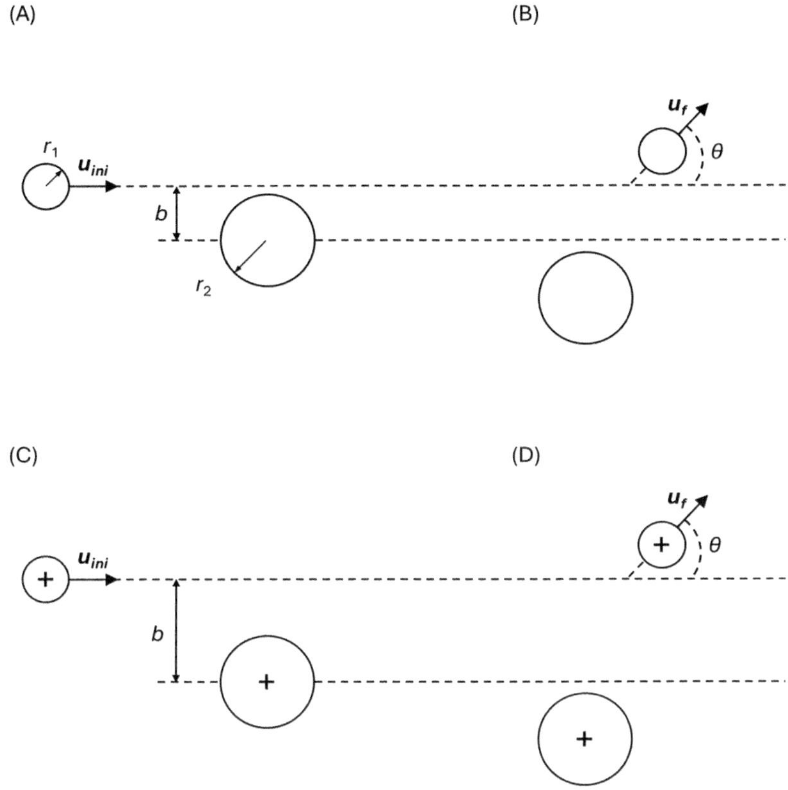

Figure 3.1 Two examples of repulsive elastic collisions. (A) The initial condition of two neutral particles with the radius r_1 and r_2 and their an initial relative velocity u_{ini}. The impact parameter $b < r_1 + r_2$. (B) The condition after their collision with a final relative velocity u_f. (C) The initial condition of two positive ions. The impact parameter $b > r_1 + r_2$ as a Coulomb collision. (D) The condition after their collision with a final relative velocity u_f.

for a collision can occur between neutral particles or one of them is charged while the other is neutral. If the particles are both charged, their long-range Coulomb force can make them repel or attract each other without a direct contact so that $b > r_1 + r_2$. The elastic Coulomb collision cross sections can be computed as[4]

$$\sigma_{e-i} = \frac{Z^2 e^4 \ln \Lambda}{4\pi\varepsilon_0^2 m^2 v_e^4} \tag{3.3a}$$

$$\sigma_{i-i} = \frac{4\pi(Z_1 Z_2 e^2)^2 \ln\Lambda}{\varepsilon_0^2 m_r^2 v_r^4} \tag{3.3b}$$

$$\ln\Lambda = 23 - \ln(n_e^{0.5} Z T_e^{-3/2}) \tag{3.3c}$$

where σ_{e-i} is the elastic electron-ion collision cross section, Z is the charge of ion, e is the unit charge, ln Λ is the Coulomb logarithm, ε_0 is the permittivity (a.k.a. dielectric constant) of free space, m is the mass of ion, v_e is the thermal velocity of the electron, n_e is the electron density, T_e is the electron temperature, σ_{i-i} is the ion-ion collision cross section with Z_1 and Z_2 representing their charge, and v_r is their relative velocity before the collision. The reduce mass is

$$m_r = \frac{m_1 m_2}{m_1 + m_2} \tag{3.3d}$$

where m_1 and m_2 are the mass of the ions. In general, the Coulomb logarithm is defined as

$$\ln\Lambda = \ln\left(\frac{\lambda_D}{b_c}\right) \tag{3.4}$$

where λ_D is the Debye length defined as

$$\lambda_D = \sqrt{\frac{\varepsilon_0 k_B T_e}{n_e e^2}} \tag{3.5}$$

The Debye length is a characteristic length scale that describes how electric fields are screened or "shielded" by the distribution of charges in plasma (Fig. 3.2). It represents the distance over which significant charge imbalances can be "felt" by a test particle before they are neutralized by the rearrangement of other charges in the plasma. Beyond this distance, the electric potential due to a charge in the plasma is exponentially reduced. Please note that the concept of Debye length is also commonly used in electrochemistry where a shielding of charge can also occur in electrolytes.

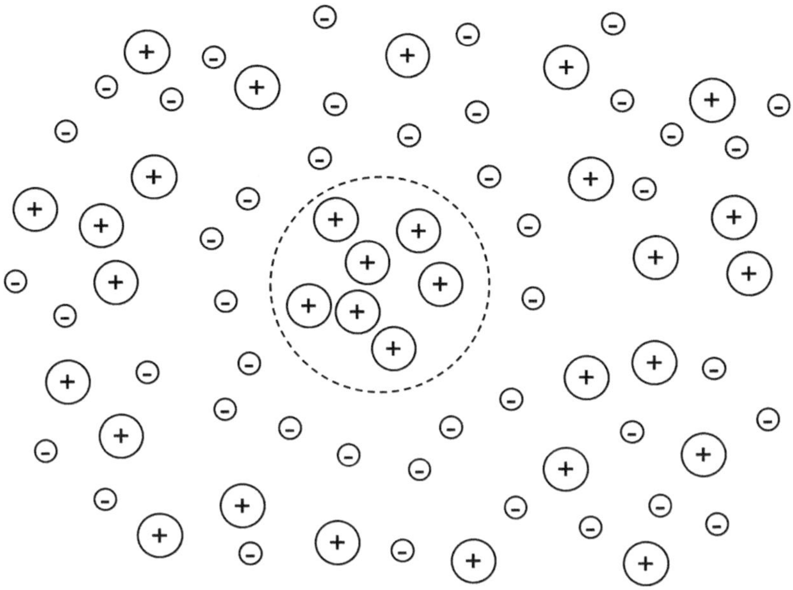

Figure 3.2 The shielding of plasma: A group of ions appears in the plasma, leading to a high concentration of electrons next to it. But these electrons also have electric fields to cancel out the field of ions in the center. Therefore, such a disturbance of quasi-neutrality has a limited range of effects. The range is described by the Debye length.

In general, the cross section for a collision between two particles depends on their relative velocity due to the nature of their interaction force. For example, in the case of Coulomb scattering (elastic scattering of charged particles), the differential cross section for a single collision at an angle θ as shown in Fig. 3.1 is given by the Rutherford scattering formula:

$$\frac{d\sigma}{d\Omega}=\left(\frac{Z_1Z_2e^2}{16\pi\varepsilon_0E_{kr}}\right)^2\frac{1}{\sin^4(\theta/2)} \tag{3.6}$$

where Ω is the solid angle that can be integrated to acquire the total cross section σ, and E_{kr} is the relative kinetic energy defined as $E_{kr} = \frac{1}{2}m_rv_r^2$. For neutral particle collisions, the collision cross section also depends on the relative velocity of the colliding particles, but the interaction is quite different from that in charged particle collisions. Since neutral particles do not experience Coulomb forces, the collisions are usually described by a potential that decreases rapidly with distance, such as the Lennard-Jones potential. However, to mathematically describe such a relationship, one needs to consider quantum physics.

3.2 Collisions in Quantum Physics Regime

As shown in Fig. 3.3, the electron orbitals are not classic orbits but actually probability of the electron locations. Therefore, a rough and general understanding can be the following: when b is small, the particles do not need high kinetic energy to reach the same scattering angle because they have high potential energy of the electron cloud. This is also the same for the inelastic collision. For example, the ionization process, when b is small, the particles have more potential energy to ionize because they are closer, rather than requiring more kinetic energy. This leads to a smaller cross section. When b is large, they need higher relative velocity to reach the same collision result. There is also another way to consider this, but actually they are the same: At higher relative velocities, particles have more kinetic energy, and they are less likely to be significantly deflected by the potential, leading to smaller cross sections. Conversely, at lower relative velocities, the cross section can increase because the particles are more easily influenced by the potential. Please note that the actual cross sections can be non-monotonic versus the velocity and the collision parameter b for specific inelastic collisions.

To quantify the collision cross section as a function of velocity in quantum physics, one needs to first consider the moving particle is a De Broglie wave, and the relative velocity can be represented by the wave number:

$$k = \frac{m_r v_r}{\hbar} \tag{3.7}$$

where k is the wave number, and h is the reduced Planck's constant.

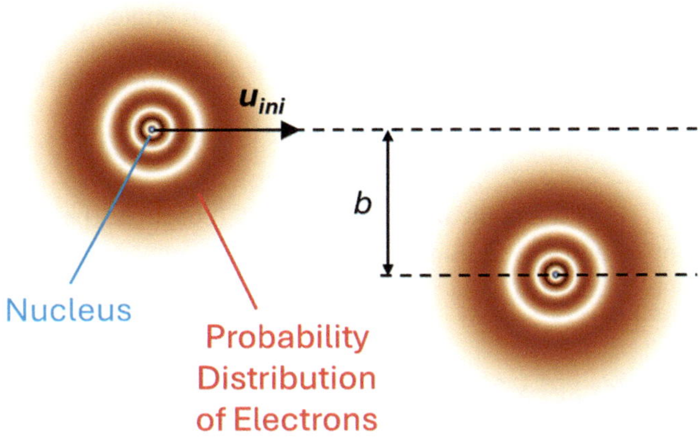

Figure 3.3 The collision of two atoms: The atomic orbitals are actually the distributions of probability to find an electron, also known as the "electron cloud". The darkness scale of the orange region is proportional to the probability, giving a rough idea of how the electron cloud can look like in an atom.

Then, the time-independent Schrödinger equation of two particles is considered to solve for the relative motion:[12,13]

$$-\frac{\hbar^2}{2m_r}\nabla^2\psi(\boldsymbol{r})+V(\boldsymbol{r})\psi(\boldsymbol{r})=E\psi(\boldsymbol{r}) \tag{3.8a}$$

$$E=\frac{(kh)^2}{2m_r} \tag{3.8b}$$

where $\psi(\boldsymbol{r})$ is the wavefunction of relative motion, $\boldsymbol{r}$ is the relative position vector, $V(\boldsymbol{r})$ is the potential, and E is the kinetic energy. The solution to the Schrödinger equation for $\psi(\boldsymbol{r})$ depending on the given the potential is spherically symmetric. Thus, we can express the Laplacian in spherical coordinates and separate the wavefunction into a radial part $R(\boldsymbol{r})$ and an angular part $Y_l^m(\theta,\varphi)$:

$$\psi(\boldsymbol{r})=R(\boldsymbol{r})Y_l^m(\theta,\varphi) \tag{3.9}$$

where l is the orbital angular momentum quantum number (non-negative integer: l = 0, 1, 2, ...), m is the magnetic quantum number (m = $-l$, $-l$, + 1, ...l), θ and φ are the polar angle and azimuthal angle that can describe the solid angle. The angular part can be computed as

$$Y_l^m(\theta,\varphi)=N_l^m P_l^m(\cos\theta)\exp(im\varphi) \tag{3.10}$$

where N_l^m is a normalization constant, $P_l^m(\cos\theta)$ is the Legendre polynomial, and i is the imaginary unit. The radial component can be solved from

$$\nabla^2 R(\boldsymbol{r})+\frac{2}{\boldsymbol{r}}\nabla R(\boldsymbol{r})+R(\boldsymbol{r})\left[k^2-\frac{l(l+1)}{r^2}-\frac{2m_r V(r)}{\hbar^2}\right]=0 \tag{3.11}$$

Combining Eq. (3.9)–(3.11), we can solve for $\psi(\boldsymbol{r})$. On the other hand, $\psi(\boldsymbol{r})$ can also be written in an asymptotic form as

$$\psi(\boldsymbol{r})\approx\exp(i\boldsymbol{k}\cdot\boldsymbol{r})+f(\theta,\varphi)\frac{\exp(ikr)}{r} \tag{3.12}$$

where $\exp(i\boldsymbol{k}\cdot\boldsymbol{r})$ representing the incident plane wave as the De Broglie wave of the incident relative motion before the collision, $\frac{\exp(ikr)}{r}$ is the outgoing spherical wave as the resulted relative motion of the collision, and $f(\theta, \varphi)$ is the scattering amplitude. Comparing Eq. (3.11) and Eq. (3.12), we can solve for $f(\theta, \varphi)$, and finally, the collision cross section in the view of quantum physics can be solved as

$$\sigma=\int|f(\theta,\varphi)|^2\,\mathrm{d}\Omega \tag{3.13}$$

One can see that the information of initial relative velocity is carried all the way to the final result of cross section in the above derivation, and it is strongly coupled with the potential $V(\boldsymbol{r})$, as we explained earlier.

3.3 Photoreactions

Photoreactions in plasmas involve various processes in which photons interact with molecules or atoms, leading to excitation, ionization, and other changes in the state of these species. These reactions are fundamental in plasma physics and have applications in fields ranging from astrophysics to semiconductor manufacturing. Photoreactions are very different from other particle collisions. Therefore, we have this standalone section to cover the excitation of molecules from the ground state, the factors determining the lifetime of excited species, the mechanisms of photon emission, and the detailed physics of photoionization and photodetachment. We will also discuss the rate coefficients and collision cross sections, with a focus on the unique considerations needed for photon-induced reactions.

In plasma, electromagnetic waves can be emitted in many ways, such as the radiofrequency (RF) emission from the plasma oscillation. However, in this chapter, we focus on the quantized photons, with the wavelength at nanometers. Therefore, the photons mentioned in this section are UV-Vis. These photons are the most commonly seen in the low-temperature plasmas which are heavily relying on the plasma chemistry in their engineering applications (see the previous chapters).

Molecules and atoms can be excited to higher energy levels through various mechanisms. One of the primary mechanisms is electron impact excitation, where a free electron collides with a molecule, transferring some of its kinetic energy to the molecule and promoting it to an excited electronic, vibrational, or rotational state. The energy required to excite a molecule from a lower state $|i\rangle$ to a higher state $|j\rangle$ can be expressed as:

$$E_{i,j} = E_j - E_i \tag{3.14}$$

where E_i and E_j are the energies of the states $|i\rangle$ and $|j\rangle$, respectively. The cross section $\sigma(E)$ for electron impact excitation depends on the energy of the incident electron and the particular transition, and it generally increases with the electron energy up to a certain point before decreasing due to the decreasing likelihood of sufficient energy transfer at very high energies. The lifetime τ of an excited species in a plasma is determined by the balance of radiative and non-radiative processes that depopulate the excited state. Radiative decay, or spontaneous emission, occurs when the excited species emits a photon and returns to a lower energy state. The rate of spontaneous emission $\frac{\partial n_{ph}}{\partial t}$ is related to the Einstein coefficient $A_{i,j}$ for spontaneous emission from state $|j\rangle$ to state $|i\rangle$:

$$\frac{\partial n_{ph}}{\partial t} = f(A_{i,j}) \tag{3.15}$$

where n_{ph} is the density of photon at a certain wavelength. The total lifetime of the excited state, considering both radiative and collisional (quenching) processes, is thus given by:

$$\tau = \frac{1}{A_{i,j} + \sum_{\zeta} n_{\zeta} Q_{j,\zeta}} \tag{3.16}$$

where n_k is the number density of colliding species ζ, and $Q_{j,\,\zeta}$ is the quenching cross section for collisions between the excited species in state $|j\rangle$ and species ζ. The total decay rate is the sum of the rates of all possible de-excitation processes.

On the other hand, when an excited molecule or atom is not quenched by the collisions, it can successfully transition to a lower energy state, it may emit a photon with an energy corresponding to the difference between the two states. The intensity $I(\lambda)$ or $I(\nu)$ of the emitted radiation at the wavelength λ or the frequency ν is proportional to the product of the population of the excited state and the Einstein coefficient for spontaneous emission:

$$I \propto N_j A_{i,j} \tag{3.17}$$

where N_j is the population density of the excited state $|j\rangle$. The emission spectrum of a plasma is thus determined by the distribution of excited states and the respective transition probabilities.

Photoionization occurs when a photon with sufficient energy interacts with an atom or molecule, causing the ejection of an electron and resulting in ionization. The minimum photon energy required for photoionization is the ionization potential of the species. The rate of photoionization is given by:

$$\frac{\partial n_{ph}}{\partial t} = n_{ph}\sigma_{ph}(h\nu)\Gamma_{ph} \tag{3.18}$$

where $\sigma_{ph}(h\nu)$ is the collision cross section as a function of the photon energy $h\nu$ and Γ_{ph} is the photon flux. Photodetachment, a similar process, involves the removal of an electron from a negatively charged ion (anion) rather than a neutral atom or molecule. The cross section for photodetachment depends on the binding energy of the electron in the ion and the photon energy. The rate coefficient for a photoreaction, such as photoionization or photodetachment, is related to the cross section σ_{ph} and the energy distribution of the incident photons, as what had been introduced in Chapter 1.

The collision cross sections of photoreactions are also different. For photon-induced reactions, such as photoionization, the collision cross section σ_{ph} is related to the absorption cross section σ_{abs} and can be expressed using quantum mechanical considerations. The absorption cross section is proportional to the oscillator strength $f_{i,j}$ of the transition and the energy of the photon:

$$\sigma_{abs} = \frac{\pi e^2}{m_e c} f_{i,j}\Phi(E_t) \tag{3.19}$$

where m_e is the electron mass, c is the speed of light, and $\Phi(E_t)$ is the line shape function centered at the transition energy E_t. For photoreactions, the cross section also incorporates the density of states of the final continuum state in the case of ionization or detachment.

3.4 The Monte-Carlo Algorithm of Chemical Pathways

In this section, we will introduce the most commonly used algorithm that is used in plasma simulation to compute plasma chemistry: Monte-Carlo collision (MCC) algorithm. It is usually combined with the particle-in-cell (PIC) to simulate plasmas, namely PIC-MCC.[14] The MCC algorithm is a powerful numerical method used to simulate and compute various phenomena in plasma chemistry, particularly when dealing with complex systems where deterministic methods may be challenging or infeasible to apply. This method relies on the use of random sampling and statistical techniques to model and understand the behavior of systems that involve numerous interacting particles, such as plasmas.

In plasma chemistry, the interactions between particles, including ions, electrons, and neutral species, are governed by a vast array of chemical reactions, collisions, and transport processes. These interactions are often highly nonlinear and can span many orders of magnitude in time and space scales. Traditional deterministic methods, like solving differential equations analytically or using finite element methods, may become impractical due to the computational complexity and the large number of variables involved.[15] This is where the MCC method becomes advantageous.[15] The MCC method in plasma chemistry is used to simulate the stochastic (random) nature of particle collisions and chemical reactions. It involves the use of random numbers to determine the outcomes of individual events, such as the probability of a collision occurring, the types of reactions that might happen, and the energy and direction of the particles after collisions.[16,17] By repeating these random events many times (often millions or billions of iterations), the MCC method can build a statistical picture of the plasma behavior, which can then be analyzed to determine macroscopic properties like reaction rates, transport coefficients, and species concentrations.

The algorithm is based on the rate equations as introduced in Chapter 1. The core of the MCC method is the random sampling of possible events that each particle can undergo during a time step. The probability of each event is determined based on cross sections or reaction rate coefficients. Let's consider a random number δ is generated between 0 and 1. A specific collision event, either an elastic collision or an inelastic one, can occur if

$$\delta \leq p \tag{3.20}$$

and

$$p = \frac{N_C}{v_{\mathrm{tot}} \mathrm{d}t} \tag{3.21}$$

where p is the probability, N_C is the number of such an event occurs in a time step (unit period of time) dt, and v_{tot} is the total collision frequency. According to the rate equation introduced in Chapter 1, N_C of a specific event can be computed as

$$N_C = \int \frac{\partial n_\zeta}{\partial t} \mathrm{d}t = \mathrm{d}t \sum_\xi \Xi_{\zeta,\xi} k_\xi \prod_{\xi,\iota} n_{\xi,\iota}^{\Delta_{\xi,\iota}} \tag{3.22}$$

where t is time, n_ζ is the density of the ζth species, $\Xi_{\zeta,\xi}$ is the coefficient of the ζth species when it is the product in the ξth reaction, k_ξ is the reaction rate coefficient of the ξth reaction that can affect the ζth species, and $n_{\xi,\iota}^{\Delta}$ is the density of the ιth reactant in the ξth reaction with a reaction degree of Δ.

Let's consider a vector $\boldsymbol{S}$ to store all the species concentrations, the size of $\boldsymbol{S}$ is thus $N_S \times 1$, where N_S is the number of species. According to the rate equation set of all the possible reactions with a total number of N_R, we have

$$\boldsymbol{S} \leftarrow \boldsymbol{S} + \boldsymbol{I}_S^T \boldsymbol{P} v_{\mathrm{tot}} \mathrm{d}t \tag{3.23}$$

where $\boldsymbol{I_S}$ is the species increment matrix with a size of $N_R \times N_S$, the superscript T represents transpose, and $\boldsymbol{P}$ is the probability vector with a size of $N_R \times 1$. The species increment matrix is defined as illustrated in Fig. 3.4

The resulting species increment matrix of the example shown in Fig. 3.4 is thus

$$\boldsymbol{I}_s = \begin{bmatrix} -1 & -1 & 1 & 0 \\ 1 & 0 & -2 & 1 \\ 1 & 2 & 0 & -1 \end{bmatrix} \tag{3.24}$$

In the first reaction, we lose 1 species A, lose 1 species B, and gain 1 species C, while species D is not involved. In the second reaction, we gain 1 species A, no species B involved, 2 species C are consumed, and gain 1 species D. In the third reaction, we acquire 1 species A, net increment 2 species B, and 1 species D is consumed with no species C involved. The probability vector is computed as

$$
\boldsymbol{P}=\frac{1}{\nu_{\text{tot}}\text{d}t}\begin{bmatrix} k_1 n_A n_B \text{d}t \\ k_2 n_C^2 \text{d}t \\ k_3 n_B n_D \text{d}t \end{bmatrix} \tag{3.25}
$$

(A)

	Species 1	Species 2	...	Species N_S
Reaction 1	Coefficient 1,1	Coefficient 2,1	...	Coefficient N_S,1
Reaction 2	Coefficient 1,2	Coefficient 2,2	...	Coefficient N_S,2
Reaction 3	Coefficient 1,3	Coefficient 2,3	...	Coefficient N_S,3
...	...	...	...	...
Reaction N_R	Coefficient 1, N_R	Coefficient 2, N_R	...	Coefficient N_S, N_R

(B)

Reaction 1: $A + B \Rightarrow C$
Reaction 2: $2C \Rightarrow A + D$
Reaction 3: $B + D \Rightarrow A + 3B$

(C)

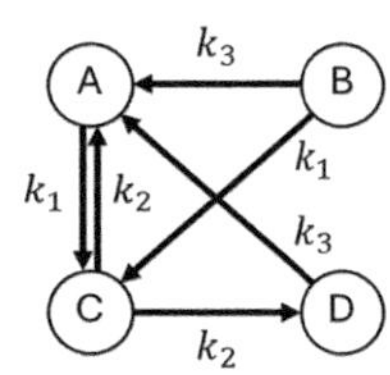

Figure 3.4 The definition of species increment matrix $\boldsymbol{I_S}$ and an example. (A) The $N_R \times N_S$ species increment matrix $\boldsymbol{I_S}$. (B) Three reactions with four species involved as an example. (C) The chemical pathway network of the example with k_1, k_2, and k_3 are the rate coefficients of these reactions.

where n_A – n_D are the density of species A to D. Combining Eq. (3.23)–(3.25), yields

$$
\boldsymbol{S}=\begin{bmatrix} n_A \\ n_B \\ n_C \\ n_D \end{bmatrix} \leftarrow \begin{bmatrix} n_A \\ n_B \\ n_C \\ n_D \end{bmatrix} + \begin{bmatrix} -1 & -1 & 1 & 0 \\ 1 & 0 & -2 & 1 \\ 1 & 2 & 0 & -1 \end{bmatrix}^T \begin{bmatrix} k_1 n_A n_B \text{d}t \\ k_2 n_C^2 \text{d}t \\ k_3 n_B n_D \text{d}t \end{bmatrix} \tag{3.26}
$$

This leads to the linear system of the set of all the rate equations agrees with what we introduced in Chapter 1.

Here is a MATLAB example of a 0-D plasma chemical simulation, and we will see how the MCC method can be used to compute a helium-air low-temperature plasma generated under a constant electric field E = 1000 V/cm from t = 0 ns to t = 20 ns with a dt = 5 ps. The chemical kinetic scheme used in this example not only contains the electrons, air species neutral particles, but also the related ions, excited states, and photons at the major wavelengths such as He (3^3S) ⇒ He (2^3P) + hν (706.52 nm). The kinetic scheme contains more than 1000 reactions including photoreactions. Refer to the appendix at the end of this book for the full table of the kinetic scheme. The electron temperature is computed based on the reduced electric field E/N where N is the total particle number, and the temperature of all other species is constant at 300 K. The initial condition is shown in Table 3.1.

Table 3.1 The initial condition of the simulation

Species	Initial Concentrations [cm^{-3}]
He	10^{18}
N_2	0.8×10^{18}
O_2	0.2×10^{18}
Others	10^6

As listed in the appendix, many of the rate coefficients are not shown as equations but computed in BOLSIG+, a well-known Boltzmann solver, which is also introduced in Chapter 1. There are 161 species included in this simulation and we cannot show all of them in a single figure. Figure 3.5 is a brief summary of some reactive chemical species commonly considered in low-temperature plasma applications, such as NO, OH, etc. introduced in Chapter 1.

The species shown in Fig. 3.5 were all at their steady states at the end of the simulation which is 30 ns. The electron density rose to a peak at almost 10^{15} cm^{-3} and then decreased to 10^{13} cm^{-3}. This is a typical overshoot of an oscillation. First, it is increased from a low initial value because of the external energy input (the electric field) which is an external stimulation of such an oscillation. The increment is playing against the electron-ion recombination and the attachments whose rates are also increased when the electron density is increased. At the moment when the electron density reached its maximum value, the drawback effects: recombination and attachments are stronger than the ionization. The electron density is finally balanced at its steady state. Meanwhile, such a non-monotonic electron density causes two periods of oscillation of NO^+ and N^+, with a small disturbance of the increment of N. In Fig. 3.6, the temporally resolved spectrum shows that the major photon emissions occur at the electron density reached its peak. Three major UVC wavelength from the helium, along with the He $(3^1D) \Rightarrow$ He (2^1P) + hν (667.82 nm), He $(3^1D) \Rightarrow$ He (2^1P) + hν (667.82 nm), and N_2^+ $(B^2\Sigma_u^+ v = 0) \Rightarrow$ N_2^+ $(X^2\Sigma_g^+ v = 1)$ + hν (391.44 nm), and N_2^+ $(B^2\Sigma_u^+ v = 0) \Rightarrow$ N_2^+ $(X^2\Sigma_g^+ v = 1)$ + hν (427.81 nm) are the most significant peaks, while N_2 $(C^3\Pi_u\ v = 0) \Rightarrow$ N_2 $(B^3\Pi_g\ v = 0)$ + hν (337.13 nm) is weaker at the moment.[18] These are all famous UV-Vis emissions for helium-air plasma.

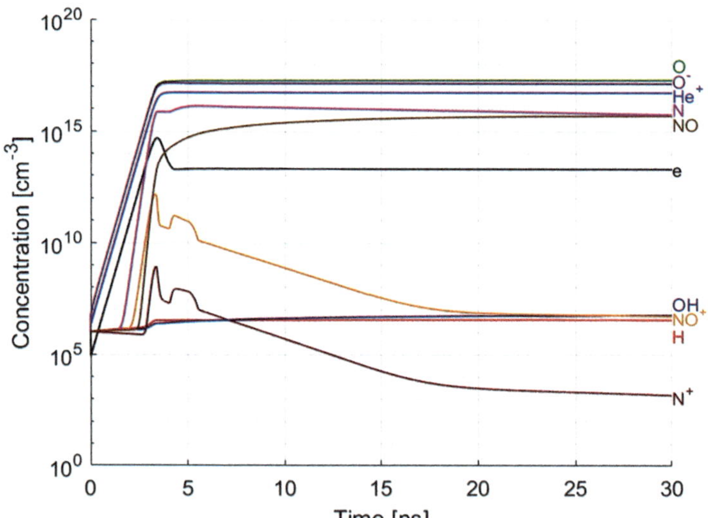

Figure 3.5 The simulation result of reactive species as an example.

The system finally reached a steady state according to the initial gas composition. If we alter the gas input, the result of steady state will surely be changed. Also, the result is determined by the external energy input, such as the electric field in this example. But such a schematic of system actually agrees with the Von Neumann architecture. In other words, if we consider the plasma system as an information processor, the initial gas composition is the input, and the external energy is the program, the whole plasma system can be a programmable computer! This also leads to a revolutionary concept of programmable intelligent material which will be discussed in the next chapter.

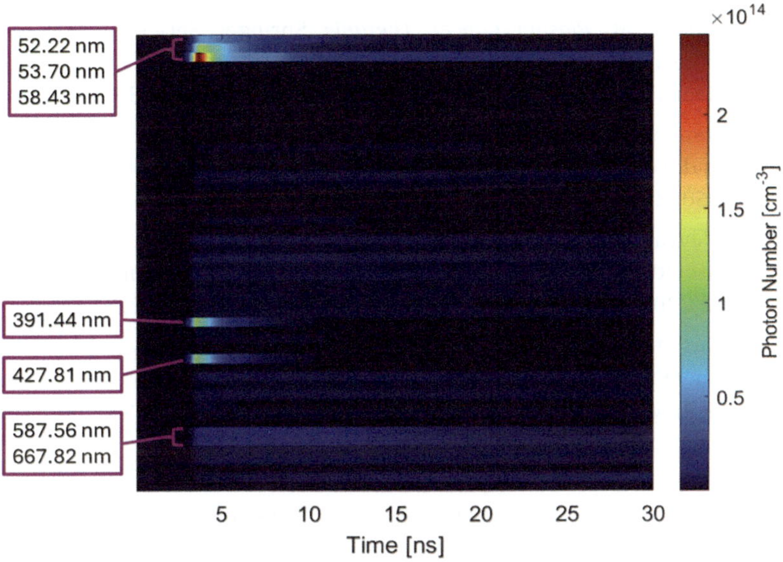

Figure 3.6 The simulation result of spectrum.

References

1. Adamovich, I.; Agarwal, S.; Ahedo, E.; Alves, L. L.; Baalrud, S.; Babaeva, N.; Bogaerts, A.; Bourdon, A.; Bruggeman, P. J.; Canal, C.; Choi, E. H.; Coulombe, S.; Donkó, Z.; Graves, D. B.; Hamaguchi, S.; Hegemann, D.; Hori, M.; Kim, H.-H.; Kroesen, G. M. W.; Kushner, M. J.; Laricchiuta, A.; Li, X.; Magin, T. E.; Mededovic Thagard, S.; Miller, V.; Murphy, A. B.; Oehrlein, G. S.; Puac, N.; Sankaran, R. M.; Samukawa, S.; Shiratani, M.; Šimek, M.; Tarasenko, N.; Terashima, K.; Thomas Jr, E.; Trieschmann, J.; Tsikata, S.; Turner, M. M.; Van Der Walt, I. J.; Van De Sanden, M. C. M.; Von Woedtke, T. The 2022 Plasma Roadmap: Low Temperature Plasma Science and Technology. *J. Phys. Appl. Phys.* **2022**, *55* (37), 373001. https://doi.org/10.1088/1361-6463/ac5e1c.

2. Fridman, A. *Plasma Chemistry*; Cambridge University Press, 2008.
3. Naidis, G. V.; Babaeva, N. Yu. Modeling of Streamer Dynamics in Helium Plasma Jets Driven by Modulated Voltage Pulses. *Phys. Plasmas* **2019**, *26* (9), 094501. https://doi.org/10.1063/1.5115779.
4. Keidar, M.; Beilis, I. *Plasma Engineering: Applications from Aerospace to Bio and Nanotechnology*; Academic Press, 2013.
5. Laroussi, M. Plasma Medicine: A Brief Introduction. *Plasma* **2018**, *1* (1), 47–60. https://doi.org/10.3390/plasma1010005.
6. Keidar, M. *Plasma Cancer Therapy*; Springer Series on Atomic, Optical, and Plasma Physics; Springer International Publishing: Cham, 2020; Vol. 115. https://doi.org/10.1007/978-3-030-49966-2.
7. Donnelly, V. M.; Kornblit, A. Plasma Etching: Yesterday, Today, and Tomorrow. *J. Vac. Sci. Technol. Vac. Surf. Films* **2013**, *31* (5), 050825. https://doi.org/10.1116/1.4819316.
8. Meyyappan, M.; Delzeit, L.; Cassell, A.; Hash, D. Carbon Nanotube Growth by PECVD: A Review. *Plasma Sources Sci. Technol.* **2003**, *12* (2), 205–216. https://doi.org/10.1088/0963-0252/12/2/312.
9. Kushner, M. J. A Model for the Discharge Kinetics and Plasma Chemistry during Plasma Enhanced Chemical Vapor Deposition of Amorphous Silicon. *J. Appl. Phys.* **1988**, *63* (8), 2532–2551. https://doi.org/10.1063/1.340989.
10. Carayannis, E. G.; Draper, J.; Iftimie, I. A. Nuclear Fusion Diffusion: Theory, Policy, Practice, and Politics Perspectives. *IEEE Trans. Eng. Manag.* **2022**, *69* (4), 1237–1251. https://doi.org/10.1109/TEM.2020.2982101.
11. Pautasso, G.; De Vries, P. C.; Humphreys, D.; Lehnen, M.; Rapson, C.; Raupp, G.; Snipes, J. A.; Treutterer, W.; Vergara-Fernandez, A.; Zabeo, L. The ITER Disruption Mitigation Trigger: Developing Its Preliminary Design. *Nucl. Fusion* **2018**, *58* (3), 036011. https://doi.org/10.1088/1741-4326/aaa137.
12. Bohm, D. *Quantum Theory*; Courier Corporation, 2012.
13. Craig, D. P.; Thirunamachandran, T. *Molecular Quantum Electrodynamics: An Introduction to Radiation-Molecule Interactions*; Courier Corporation, 1998.
14. Birdsall, C. K. Particle-in-Cell Charged-Particle Simulations, plus Monte Carlo Collisions with Neutral Atoms, PIC-MCC. *IEEE Trans. Plasma Sci.* **1991**, *19* (2), 65–85. https://doi.org/10.1109/27.106800.

15. Kroese, D. P.; Brereton, T.; Taimre, T.; Botev, Z. I. Why the Monte Carlo Method Is so Important Today. *WIREs Comput. Stat.* **2014**, *6* (6), 386–392. https://doi.org/10.1002/wics.1314.
16. Rubinstein, R. Y.; Kroese, D. P. *Simulation and the Monte Carlo Method*; John Wiley & Sons, 2016.
17. James, F. Monte Carlo Theory and Practice. *Rep. Prog. Phys.* **1980**, *43* (9), 1145.
18. Hagelaar, G. J. M.; Pitchford, L. C. Solving the Boltzmann Equation to Obtain Electron Transport Coefficients and Rate Coefficients for Fluid Models. *Plasma Sources Sci. Technol.* **2005**, *14* (4), 722–733. https://doi.org/10.1088/0963-0252/14/4/011.

Chapter 4

The Intelligent Plasma

Li Lin and Michael Keidar

School of Engineering and Applied Science, The George Washington University, Washington, DC 20052, USA

lilin@email.gwu.edu

4.1 The Mathematical Similarity

In Chapter 1 and Chapter 3, we have learned that the rate of change of a species concentration is a summation of the rate of all the reactions it involved. For example, let's consider a chemical pathway network (CPN) as shown in Fig. 4.1, which represents a system of two reactions: $2C + O_2 \Rightarrow 2CO$ and $C + O_2 \Rightarrow CO_2$. This is what happens when carbon is burning in the air. Their rate coefficients can be estimated as:

$$k_1 = 3 \times 10^{12} \exp\left(-\frac{1.51 \times 10^5}{RT}\right) \tag{4.1a}$$

$$k_2 = 10^{12} \exp\left(-\frac{2.19 \times 10^5}{RT}\right) \tag{4.1b}$$

Machine Learning Plasmas and the Neuromorphic Plasma Chemistry
Edited by Li Lin and Michael Keidar

ISBN 978-981-5129-84-7 (Hardcover), 978-1-003-71013-4 (eBook)
www.jennystanford.com

DOI: 10.1201/9781003710134-4

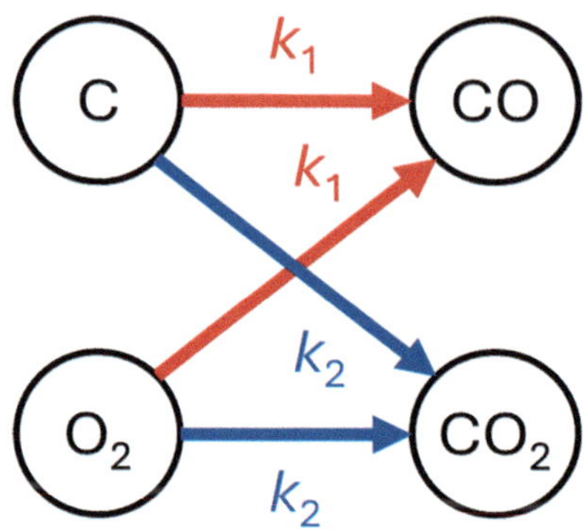

Figure 4.1 An example of a two-reaction system.

where k_1 is the rate coefficient of generating CO, k_2 is the value of generating CO_2, R is the ideal gas constant, and T is the temperature. If the system has a continuous and constant input of carbon and oxygen, along with a constant temperature during the burning, it should be able to output constant product composition. The ratio of CO and CO_2 in the output depends on two factors together: (1) the input carbon and O_2 ratio and (2) the temperature. Therefore, a Von Neumann architecture is hidden there (see Fig. 4.2) if we consider the C + O_2 chemical composition as the input, the temperature control is the program, and the CO + CO_2 mixture is the output. Moreover, we also have a similar topology of an artificial neural network (ANN) where the matrices of weights and bias can be the program, along with the input and output vectors as shown in Fig. 4.3.

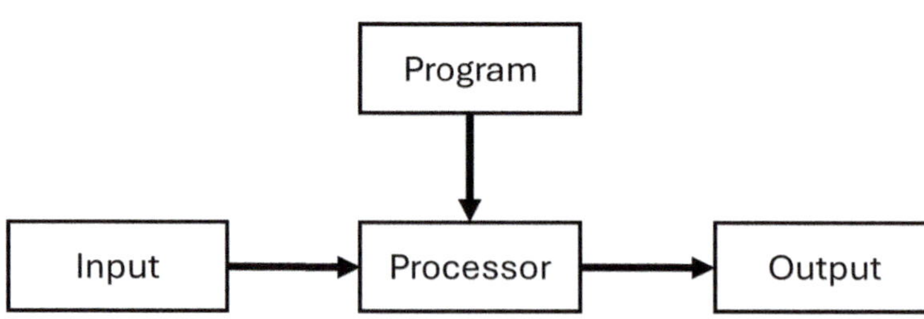

Figure 4.2 The Von Neumann architecture known as the fundamental structure of modern computers.

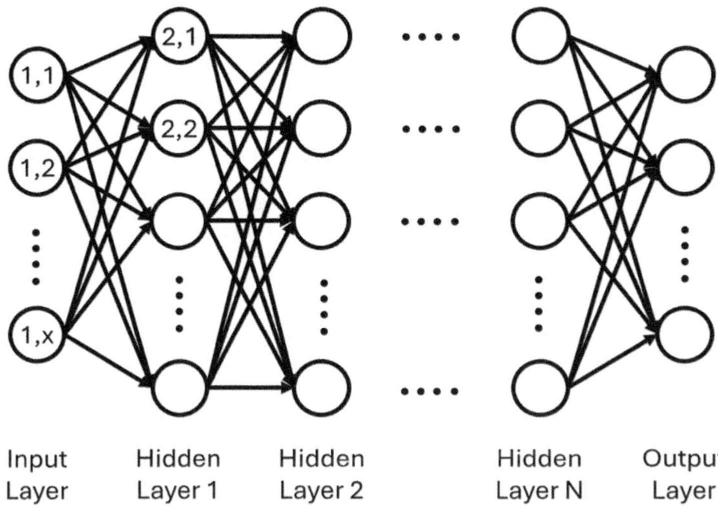

Figure 4.3 A typical example of ANN: The fully connected ANN.

Can we consider a CPN as an ANN, or even use a chemical system as a programmable material which can potentially acquire artificial intelligence? We will now find their mathematical similarity. According to what we have introduced in Chapter 1 and Chapter 3, we have the following rate equations to compute the output of a CPN:

$$\frac{\partial n_\zeta}{\partial t} = \sum_\xi \Xi_{\zeta,\xi} k_\xi \prod_{\xi,\iota} n_{\xi,\iota}^{\Delta_{\xi,\iota}} \tag{4.2}$$

where t is time, n_ζ is the density of the ζth species, $\Xi_{\zeta,\xi}$ is the coefficient of the ζth species when it is the product in the ξth reaction, k_ξ is the reaction rate coefficient of the ξth reaction that can affect the ζth species, and $n_{\xi,\iota}^{\Delta_{\xi,\iota}}$ is the density of the ιth reactant in the ξth reaction with a reaction degree of $\Delta_{\xi,\iota}$. According to Chapter 2, we have the following equation to compute the output of an ANN:

$$\boldsymbol{h}^{(l)} = \alpha[\boldsymbol{w}^{(l)} \boldsymbol{h}^{(l-1)} + \boldsymbol{b}^{(l)}] \tag{4.3}$$

where $\boldsymbol{w}$ is a set of matrices of weights, $\boldsymbol{b}$ is a set of matrices of bias, $\boldsymbol{h}^{(l)}$ is the vector of a neuron layer l, and α is the activation function. The similarity is based on the analogy shown in Table 4.1.

Table 4.1 The analogy between CPN and ANN

CPN	ANN
Species Concentrations	Neurons
Species Temperatures	Weights
Rate Coefficients	Activation Functions
Input Composition	Input Vector
Output or Diagnostics on Compositions	Output Vector

First, it is obvious that the concentration of each species is an analogy of a neuron in an ANN, that both are variables. Therefore, the complexity of the CPN depends on the number of species, which determines the fitting capability of the CPN (will be discussed in Section 4.2). The input and output of these two networks are also obviously similar: adding chemical species into a CPN is equivalent to input a vector to an ANN, while measuring some of the species' concentrations can acquire the information, equivalent to getting the output vector from an ANN. However, it is also obvious that the CPN in a chemical system is a recurrent neural network (RNN) because even some of the products of reactions cannot be the reactants of other reactions, the particles can still collide with others, leading to kinetic energy transfers. Also, it is worth emphasizing that in plasma, each species may have a different temperature. Especially for the ions that each of them can have a unique mass-to-charge ratio which can be heated using external electromagnetic spectrum (will be discussed in Section 4.3 of hardware approach). But before we discuss energy transfer, which should be an analogy of transfer of weights in ANN, let's compare the rate coefficient function in Eq. (4.2) and the activation function in Eq. (4.3), which are shown in Fig. 4.4 (The mathematical expressions of these activation functions can be found in Chapter 2). For both rate coefficient function and the activation functions, the common main feature is a stair-like increase: when the input is higher than a threshold, the output starts to increase significantly. This is so called "activation". In chemistry, this is the threshold of a chemical reaction to occur. Based on the collision theory we introduced in Chapter 3; the threshold is a requirement of the kinetic energy of the relative velocity of the particles before their collision.

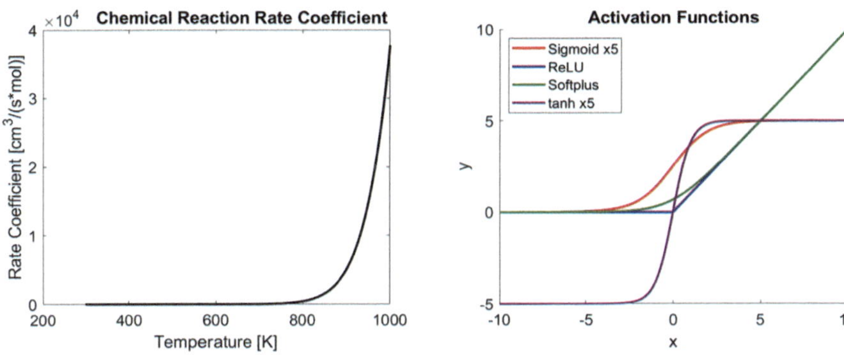

Figure 4.4 The rate coefficient function of CPNs versus the commonly used activation functions of ANNs. Here we are using the Arrhenius equation as an example of the rate coefficient function.

However, we can also see a major difference between them: the activation functions (the rate coefficient functions) of CPNs are applied on the weights (temperatures) only, rather than the weighted linear combination of neurons. While this is similar to the self-modulation of ANN, it is still a very unique network design, because none of the dynamic weight techniques in machine learning (ML) use activation functions on weights without neurons. We will have a deeper discussion on this unique CPN architecture in the next section.

4.2 Weight-Activated Recurrent Network

In view of network, Eq. (4.2) can be modified by using matrix operators instead of linear combination, with the input and output of gases also considered to the chemical system, yields

$$\boldsymbol{S}_{t+\mathrm{d}t} = \kappa(\boldsymbol{T}_t)\boldsymbol{R}_t(\boldsymbol{S}_t) + \Delta\boldsymbol{S}_t \tag{4.4}$$

where $\boldsymbol{S}$ is the vector of species concentrations, $\boldsymbol{t}$ is time, $\boldsymbol{\kappa}$ is the activation function, $\boldsymbol{T}$ is the vector of species temperature (only ions and the electron can have unique temperatures, while all other neutral species may share the same value), $\boldsymbol{R}$ is a function of $\boldsymbol{S}$ which outputs the reactant products, and $\Delta\boldsymbol{S}_t$ is the net input of the gases. The general topology of a CPN can be shown in Fig. 4.5, where each connection, rather than the weighted linear combination, is controlled by the activation function. Also, since all the neurons are connected, the network is recurrent as mentioned. Just like all the RNN, a chemical system is time dependent. Therefore, we can name this novel and unique network architecture as Weight-Activated Recurrent Network (WARN).

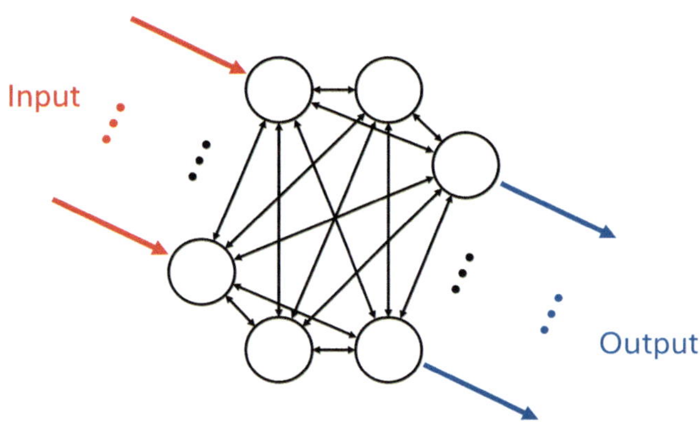

Figure 4.5 The general topology of a CPN.

The input sensitivity of a neuron's output $\boldsymbol{S}_{t+dt}$ with respect to an input $\boldsymbol{S}_t$ is governed by the derivative of the neuron function $\boldsymbol{R}_t(\boldsymbol{S}_t)$ and the activation function applied to the weight $\kappa(\boldsymbol{T}_t)$. As we can see in Fig. 4.4, small changes in $\boldsymbol{T}_t$ could significantly affect the value of $\kappa(\boldsymbol{T}_t)$. We can quantify the sensitivity of the network:

$$\frac{\partial S_{i,t+\mathrm{d}t}}{\partial S_{j,t}} = \kappa_{i,j}(T_{i,j,t})\frac{\partial}{\partial S_{j,t}}R_{i,j,t}(\boldsymbol{S}_t) \tag{4.5}$$

where the subscripts i and j are the indexes of species. The weight sensitivity depends on the derivative of $\kappa(\boldsymbol{T}_t)$ with respect to $\boldsymbol{T}_t$ as

$$\frac{\partial S_{i,t+\mathrm{d}t}}{\partial T_{i,j,t}}=R_{i,j,t}(\boldsymbol{S}_t)\frac{\partial}{\partial T_{i,j,t}}\kappa_{i,j}(T_{i,j,t}) \tag{4.6}$$

Again, since the activation function is sensitive to the weights, even small adjustments during training could lead to significant changes in the neuron outputs.

Here we can see that the weights themselves determine activations, potentially allowing for unique and complex dynamic behaviors. Neuron values also determine sensitivity which is also a common feature for all other conventional ANNs. However, the input $\Delta\boldsymbol{S}_t$ is not in the derivative (sensitivity) equations, it comes at the cost of increased training complexity and potential sensitivity to parameter initialization and learning rates. If we are using error gradient descent method to train the CPN, the WARN architecture of it may have similar performance compared with other conventional ANNs. Suppose the error at time step τ is given by a loss function $L(t)$ that depends on the network's output at that specific time step, yields

$$L(t)=\frac{1}{2}\sum_j[y_j(t)-x_j(t)]^2 \tag{4.7}$$

where $x_j(t)$ is the output of neuron j and $y_j(t)$ is the target value of it at the time τ. The update rule for neuron j at the next time step as:

$$x_j(t+1)=\sum_{i\neq j}\kappa_{i,j}(T_{i,j,t})R_{i,j,t}(\boldsymbol{S}_t)+S_{j,t} \tag{4.8}$$

If there is no input on neuron j, the last term on the right-hand side in Eq. (4.8) will be zero. Based on these two equations, we can derive the backpropagation equation for the WARN architecture. First, compute the partial derivatives

$$\frac{\partial L(t)}{\partial x_j(t)} = -y_j(t) + x_j(t) \tag{4.9a}$$

$$\frac{\partial x_j(t+1)}{\partial T_{i,j,t}} = R_{i,j,t}(\mathbf{S}_t) \frac{\partial}{\partial T_{i,j,t}} \kappa_{j,t}(T_{i,j,t}) \tag{4.9b}$$

Then, using the chain rule of derivation, the gradient of the error with respect to weight at the time τ becomes

$$\frac{\partial L(\tau)}{\partial T_{i,j,t}} = \sum_{t=0}^{\tau} R_{i,j,t}(\mathbf{S}_t) \frac{\partial}{\partial T_{i,j,t}} \kappa_{j,t}(T_{i,j,t})[-y_j(t) + x_j(t)] \tag{4.10}$$

One of the advantages of CPN is that the system is immune to the error gradient vanishing and exploding. The error gradient vanishes when $\frac{\partial}{\partial T_{i,j,t}} \kappa_{i,j}(T_{i,j,t}) \ll 1$. As shown in Fig. 4.4, the slope of rate coefficient function $\kappa_{i,j}(T_{i,j,t})$ becomes low only when the temperature is low. However, because of the WARN architecture of CPNs, the temperature can be directly controlled externally, the error gradient vanishing is also under control. The explosion is obviously also not an issue, because of the upper limits of the temperatures.

The next performance that must be discussed is the fitting capability of CAP. Apparently, like other ANNs, the fitting capability depends on the complexity of the CPN. In other words, the more species included in a CPN, the more neurons and more connections are contained in its WARN architecture. However, a feature of a chemical system, like plasma, is that some of the species' concentrations may have fixed relationships. For example, when quenching collisions are at very low frequency, the emission rate of a specific wavelength photon becomes linear depending on the density of an excited species and depending on it only. This is because of the constant lifetime of an excited state. These fixed relationships are not variables, and they may lead to a decrement of complexity, and so is the fitting capability. However, it can easily acquire massive number of neurons (species) in a chemical system, considering each atom and molecule can have multiple excited states, and their combinations can make more states. All these excited states are different neurons in the network.

A serious quantification and discussions on this topic have been published a few years ago.[1] The paper shows that a helium-air plasma can reach an artificial intelligence level high enough to play a board game.[1] The CPN in this work was considered as a helium-air plasma at a pressure slightly lower than the atmospheric pressure to achieve a balance between the ion cyclotron resonance efficiency and the collision frequency.[1] The 109 chemical species with 905 reactions, including the photoreactions with numerous different wavelengths, ensured a complexity of the network that can achieve an AI enough to play a board game Tic-Tac-Toe.[1] A schematic of how such an intelligent plasma can play it is shown in Fig. 4.6. First, nine different gases will represent the nine locations on the board. If there is a circle at a board location, the corresponding gas concentration will be increased by multiplying a constant φ, and if there is a cross at that location, the corresponding gas concentration will be decreased by a factor of φ. Therefore, each and every possible board status can be represented by a unique chemical composition of a gas mixture. The gas mixture will thus be input to the plasma reactor. The plasma will be generated under a set of constant discharge parameters such as the discharge voltage, frequency, and the external RF spectrum with multiple peaks to control the ion temperatures. The plasma will then output a set of UV-Vis spectra representing the plasma's next move. Therefore, the goal of training the plasma to achieve intelligence to play the board game is to find the set of constant discharge parameters in the reactor that no matter what kind of gas input (board status), the plasma can always output a correct UV-Vis spectra (the best next move). Please note that the discharge parameters are not variables, they are constant. Therefore, the plasma's next move is not controlled by these discharge parameters, otherwise, it will be the system controlling the plasma playing the game rather than the plasma playing by itself. The authors finally achieved a significant winning rate of their plasma playing against a random-move player.[1] Intelligent plasma can win almost 90% of plasma's first-move games, and more than 50% of the opponent's first-move games. The average winning rate is more than 70% of all the games it played.

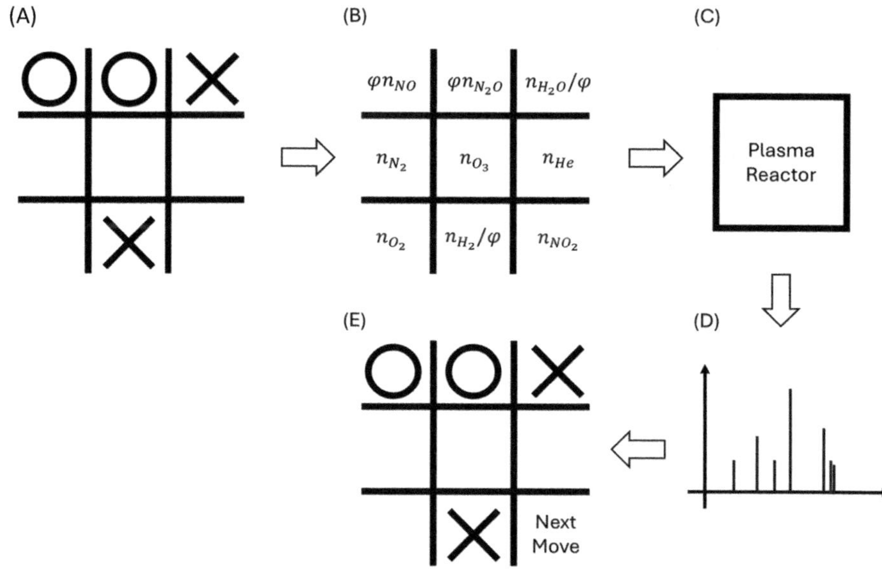

Figure 4.6 The schematic of an intelligent plasma playing Tic-Tac-Toe.[1] (A) The current board status. (B) Each location on the board is represented by a concentration of gas. Each possible board status is thus represented by a unique gas mixture input to the plasma reactor. (C) The plasma is generated in a reactor with a set of constant discharge parameters. (D) The plasma outputs a set of spectra. (E) The spectra will represent what the plasma's next move will be.

The intelligent plasma can even discover and develop its own strategy of playing the game, such as creating two ways of winning, so that the opponent can only block one of them, to ensure the victory.[1] After a statistical study of the plasma's playing style over the training, the authors found that the intelligent plasma became more and more aggressive during the training by winning a game with less moves. However, it is surprising that the authors never setup an award on winning with less moves.[1] We thus can conclude that plasma became aggressive by itself. The authors also quantified the complexity of CPN in intelligent plasma as a function of species numbers in the system versus the missions it can possibly accomplish.[1] However, this is obviously just the beginning of such a study, and future deeper works and analysis are required to find the full potential of such a theory. In the next section, we will discuss the hardware that can be used to achieve such material-based intelligence.

4.3 Hardware Approaches

As mentioned in Section 4.1, to enable a CPN, one needs to achieve an effective control on the ion temperatures which is an analogy to the weights in ANNs. In this section, we will discuss some of the possible hardware approaches.

The first one is to use Electromagnetic Spectrum Targeting (EST). The principle behind this approach is to exploit the resonant frequencies of ions. Each ion species has a unique set of resonant frequencies at which it absorbs energy most efficiently. By tuning an external electromagnetic field to these specific frequencies, the user can selectively heat the desired ion species. This method leverages the fact that ions with different mass-to-charge ratios will have different resonant frequencies. This method leverages the principle of resonance heating, where ions absorb energy most efficiently at frequencies corresponding to their natural cyclotron or Larmor frequencies. By tuning the external electromagnetic field to these specific frequencies, the ions with the targeted mass-to-charge ratio will resonate and absorb more energy, leading to selective heating. In more detail, the cyclotron frequency of an ion in a magnetic field is given by:

$$\omega_c = 2\pi v_i = \frac{qB}{m_i} \tag{4.11}$$

where q is the charge of ion with a mass of m_i exposed in an external magnetic field of B. By adjusting the frequency of the electromagnetic waves to match the cyclotron frequency of a specific ion species, one can selectively heat that species while minimally affecting others, as ions with different mass-to-charge ratios will not resonate at this frequency. Also, please note that the magnetic field contains not only one frequency but multiple frequencies to aim at multiple ion species. A plasma reactor could be equipped with the microwave or other radiofrequency (RF) sources, tuned to the cyclotron frequencies of the ions that are targeted. Furthermore, the amplitude and phase of these electromagnetic waves can be adjusted to optimize the energy transfer to the specific ion species, enhancing the selectivity of the heating process.

Another upgrade of EST is to modulate the magnetic field within the plasma. By varying the magnetic field strength in a controlled manner, the cyclotron frequency of the ions can be dynamically shifted to match the frequencies of an external RF source. This method allows for more flexible control over which ions are heated, as different ions can be targeted sequentially by adjusting both the magnetic field and RF frequencies. This approach requires precise synchronization between the magnetic field modulation and RF sources but could offer high selectivity. Please note that although EST is based on the ion cyclotron resonance, it does not require a full cycle of ions. In other words, EST can be achieved by heating for high pressure, in other words, high collision frequency plasmas. Therefore, it is not exactly the same as Ion Cyclotron Resonance Heating (ICRH) which is commonly used in fusion reactors.[2,3]

Another hardware approach to control the specific ion temperature is laser Doppler-shifted resonance heating.[4] In this approach, a laser beam is tuned to a frequency slightly above or below a specific ionic transition.[4] Due to the Doppler effect, ions moving toward or away from the laser will see a frequency shift, and only those ions with velocities that shift the laser into resonance will absorb the energy. This can selectively heat ions based on their velocity distribution, which indirectly depends on their mass-to-charge ratio. The Doppler effect describes the change in frequency of a wave in relation to an observer moving relative to the source of the wave. For ions in a plasma, this effect means that the frequency of incoming laser light will appear shifted depending on the velocity of the ions. If an ion is moving toward the laser source, the frequency of the laser light will appear higher (blue-shifted) to the ion. Conversely, if the ion is moving away from the laser source, the frequency will appear lower (red-shifted). For ions with a specific velocity, there exists a resonance condition where the Doppler-shifted laser frequency matches an electronic frequency, vibrational transition frequency, or the cyclotron frequency of the ion. When this condition is met, the ions can efficiently absorb the laser energy, leading to selective heating of that ion population. Specifically, let's consider a laser beam interacts with moving ions in plasma, the frequency of the laser light experienced by the ions is shifted due to the Doppler effect. For an ion moving with velocity relative to the laser source, the observed frequency υ' can be computed as

$$\upsilon' = \upsilon_0 \left(1 + \frac{v \cdot \hat{k}}{c} \right) \tag{4.12}$$

where υ_0 is the original frequency of the laser beam with a unit vector k, υ is the ion velocity, and c is the lightspeed. When a laser beam is applied on plasma, the total momentum change for a specific ion is

$$\Delta p = N_i N_{ph} \frac{h\upsilon_i}{c} \tag{4.13}$$

where Δp is the change of momentum, N_i is the number of ions, N_{ph} is the number of photons absorbed, h is the Planck's constant, and υ_i is the ion frequency. Let's consider the ion frequency is the cyclotron frequency as defined in Eq. (4.11). With the resonance condition, we have

$$\upsilon = \upsilon_i \tag{4.14}$$

Next, the kinetic energy change of an ion due to the absorbed photons can be calculated using the change in momentum as

$$\Delta E_k = \frac{(p + \Delta p)^2}{2m_i} - \frac{p^2}{2m_i} \tag{4.15}$$

where p is the original momentum, and ΔE_k is the change of kinetic energy of ions. Combining Eq. (4.13), Eq. (4.14), and Eq. (4.15), yields

$$\Delta E_k = \frac{N_i N_{ph} h\upsilon_0 v}{c} + \frac{N_i^2 N_{ph}^2 h^2 \upsilon_0^2}{2m_i c} \tag{4.16}$$

Using Eq. (1.1), we can find the ion temperature change ΔT_i as

$$\Delta T_i = \frac{2}{f k_B}\left(\frac{N_i N_{ph} h\upsilon_0 v}{c} + \frac{N_i^2 N_{ph}^2 h^2 \upsilon_0^2}{2m_i c} \right) \tag{4.17}$$

where f is the degree of freedom of the ion particle, and k_B is the Boltzmann constant. Please also note that in the experiment, it is more practical to use the change of specific intensity ΔI in the laser spectrum instead of using N_{ph}. These two parameters can be transferred by

$$N_{ph}(\lambda) = \frac{\Delta I(\lambda)\sigma(\lambda)}{h\upsilon_0} \tag{4.18a}$$

$$\lambda = \frac{c}{\upsilon_0} \tag{4.18b}$$

where σ is the absorption cross section as a function of the wavelength λ of the laser.

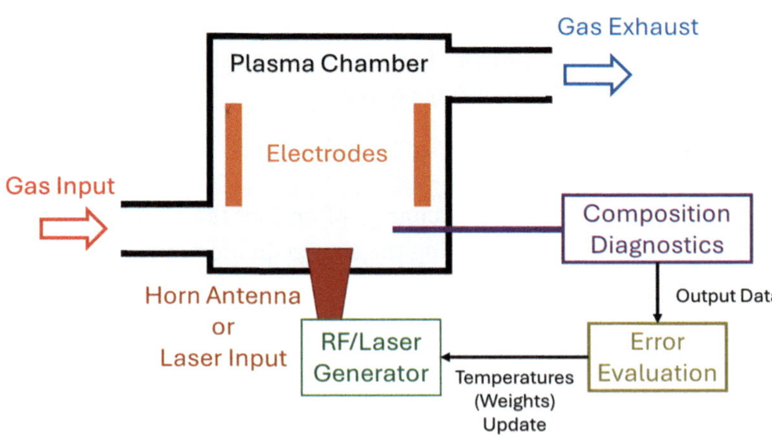

Figure 4.7 The training of CPN in an intelligent plasma.

Therefore, there are many hardware approaches to manipulate the specific ion temperatures. These temperatures in a CPN are the analogies of weights in an ANN. Therefore, in the training of CPN, the user must update these temperatures of plasma as shown in Fig. 4.7.

4.4 Supramolecules and Nanomachines

Supramolecules are large complexes formed from the assembly of smaller molecular entities through non-covalent interactions, such as hydrogen bonding, van der Waals forces, electrostatic interactions, and π–π stacking.[5–9] Unlike traditional molecules where atoms are held together by covalent bonds, supramolecules leverage these weaker forces to form highly organized structures. The field of supramolecular chemistry emerged in the late 20th century, building on the principles of molecular recognition and self-assembly. It has become a critical area of research due to its relevance in creating complex, functional materials and systems, such as mimicking biological processes.[10,11]

In the view of physical chemistry, the self-assembly process of supramolecules can often be described using thermodynamic principles. The formation of a supramolecular complex is generally driven by the minimization of Gibbs free energy ΔG:

$$\Delta G = \Delta H - T\Delta S \tag{4.19}$$

where ΔH is the change of enthalpy, T is the temperature, and ΔS is the change of entropy. For supramolecular assembly to occur spontaneously, ΔG must be negative. The balance between enthalpic gains (such as hydrogen bonding and electrostatic interactions) and entropic losses (due to the ordering of molecules) dictates the stability and structure of the supramolecular assembly.

Rotaxanes and catenanes are classic examples of mechanically interlocked supramolecular structures (see Fig. 4.8). Rotaxanes consist of a dumbbell-shaped molecule threaded through a macrocycle, with bulky end groups that prevent the ring from slipping off. The structure can be thought of as a molecular axle and wheel, where the ring (macrocycle) can move along the axle, allowing for controlled motion at the nanoscale. This motion is particularly useful in creating molecular switches and shuttles. Catenanes are composed of two or more interlocked macrocyclic rings that cannot be separated without breaking covalent bonds. The intertwined rings of catenanes can rotate relative to one another, which has been exploited to develop molecular motors and pumps that operate on chemical or photonic stimuli. These supramolecular structures have promising applications in nanomechanics and molecular electronics. For example, rotaxanes have been used as molecular switches in electronic devices. A notable example is the development of molecular electronic circuits, where the shuttling motion of the macrocycle along the axle in response to a chemical stimulus can be used to switch between conductive and non-conductive states, effectively creating a nanoscale switch. Catenanes have been used to construct molecular motors that harness light or chemical energy to drive directional rotation, a key concept for developing molecular machines that mimic the functions of biological motors, like those involved in muscle contraction.

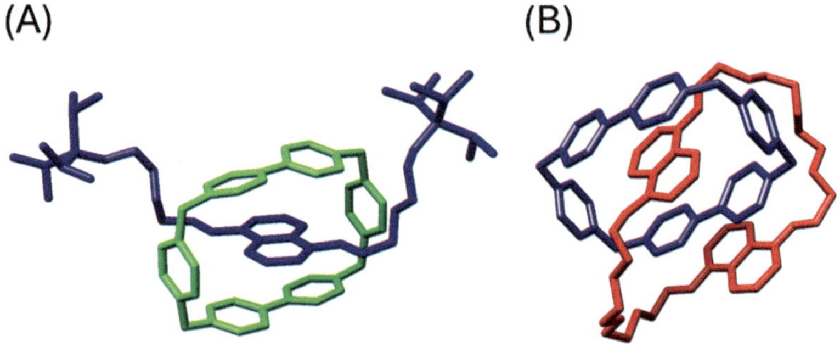

Figure 4.8 The most two iconic supramolecules: (A) rotaxanes and (B) catenanes. The subplots are reproduced under the permission of CC BY-SA 3.0 from the Wikipedia pages of "Rotaxane" and "Catenane".

Supramolecular capsules and cages are structures that can encapsulate guest molecules within a defined cavity. These structures are formed by the self-assembly of multiple building blocks, often driven by non-covalent interactions like hydrogen bonds, metal coordination, or van der Waals forces. The resulting cavities can be tailored to host specific guest molecules, allowing for highly selective interactions. Supramolecular capsules and cages are particularly useful in catalysis and drug delivery. In catalysis, these structures can mimic the function of enzymes by creating a microenvironment that stabilizes transition states or pre-organizes reactants, thereby enhancing reaction rates and selectivity. For instance, metal-organic cages have been used to encapsulate reactive intermediates, protecting them from side reactions and allowing for highly controlled catalytic processes. In drug delivery, supramolecular capsules can encapsulate therapeutic agents and protect them from degradation in the biological environment, releasing the drug in response to specific stimuli, such as pH changes or redox conditions. The supramolecular capsules[12] will be discussed in detail in the next section.

Another type of interesting example is the recent advancements in supramolecular chemistry have led to the development of structures that act like hinges and latches at the molecular level. These supramolecular constructs are often built using metal-coordinated frameworks or peptide-based designs, which can undergo conformational changes in response to external stimuli, such as light, pH, or ion concentration. Supramolecular hinges are structures that can open or close, much like a mechanical hinge, in response to specific stimuli.[13] One example includes metal-organic frameworks (MOFs) that change their conformation upon binding to certain ions. This property can be utilized to create molecular sensors or gatekeepers that control access to the inner cavity of a structure, only opening when a specific molecule is present.[13] Supramolecular latches are mechanisms that can lock a structure in place or release it upon receiving a specific signal.[13] An example is a peptide-based latch, where hydrogen bonding interactions can be disrupted under certain conditions (e.g., changes in pH or the presence of a specific ligand), causing the latch to open.[13] These latches can be used to control the release of encapsulated molecules, making them valuable in targeted drug delivery systems where the release of the therapeutic agent needs to be tightly controlled.[13] These dynamic structures have applications in smart materials and responsive nanodevices. For example, molecular hinges and latches can be used to design smart coatings that respond to environmental changes, such as humidity or temperature, by altering their properties (e.g., becoming more or less permeable). They can also be incorporated into molecular robotics, where sequences of conformational changes, triggered by external stimuli, can perform complex tasks such as transporting cargo at the nanoscale.

Nanomachines, a.k.a. nanorobots, are molecular-scale devices that perform mechanical or electronic functions at the nanoscale, typically involving components like molecular motors, switches, and rotors.[14] The field of nanomachines owes much of its conceptual foundation to Richard Feynman's 1959 lecture, "There's Plenty of Room at the Bottom", and has since been propelled by advancements in molecular biology, chemistry, and physics.[15] Nanomachines operate by converting energy into mechanical motion at the molecular level.[13] One of the fundamental physical laws governing the operation of nanomachines is Brownian motion, which describes the random motion of particles suspended in a fluid due to collisions with molecules of the fluid. For nanomachines to perform controlled movement or work, they must operate within this stochastic environment, often using energy sources like chemical reactions, light, or electrical potential to bias the motion in a specific direction. A simple model for a nanomachine, such as a molecular motor that can be described using Langevin dynamics, where the equation of 1D motion for a particle with mass under the influence of a potential and friction can be computed as:

$$m\frac{\mathrm{d}^2 x}{\mathrm{d}t^2} = -\gamma\frac{\mathrm{d}x}{\mathrm{d}t} - \frac{\mathrm{d}U(x)}{\mathrm{d}x} + \sqrt{2\gamma k_B T}\,\chi(t) \tag{4.20}$$

where m is the mass of particle, x is the displacement, γ is the friction, U is the potential, and $\chi(t)$ is a random noise term with zero mean and unit variance. The term $\sqrt{2\gamma k_B T}\chi(t)$ is a stochastic term representing thermal fluctuations.

A good example of molecular motor is the bacterial flagellar motor[16] as shown in Fig. 4.9(A). The motor can reach up to 1000 rpm with the flagellar and the rotor can reach up to 100000 rpm without the flagellar. The bacterial flagellar motor is a fascinating example of a molecular motor that converts chemical energy into mechanical work, allowing bacteria to swim through their environment. This complex rotary nanomachine is embedded in the cell membrane and consists of multiple protein components that work together to rotate the flagellum, a whip-like appendage, enabling bacterial motility. FliG, FliM, and FliN proteins are essential components of the C-ring, a cytoplasmic ring structure located on the inside of the bacterial inner membrane, the ring is the rotor that receive the torque from MotAB.[16] The MS-ring (Membrane and Supramembrane ring) is a structural component embedded within the inner membrane of the bacterium.[16] It serves as the foundation upon which other motor proteins are assembled. The MS-ring is primarily composed of FliF proteins and provides a scaffold for the assembly of the C-ring and the rotor complex.[16] It acts as the anchor point for the flagellum and is central to the overall stability and integrity of the flagellar motor.

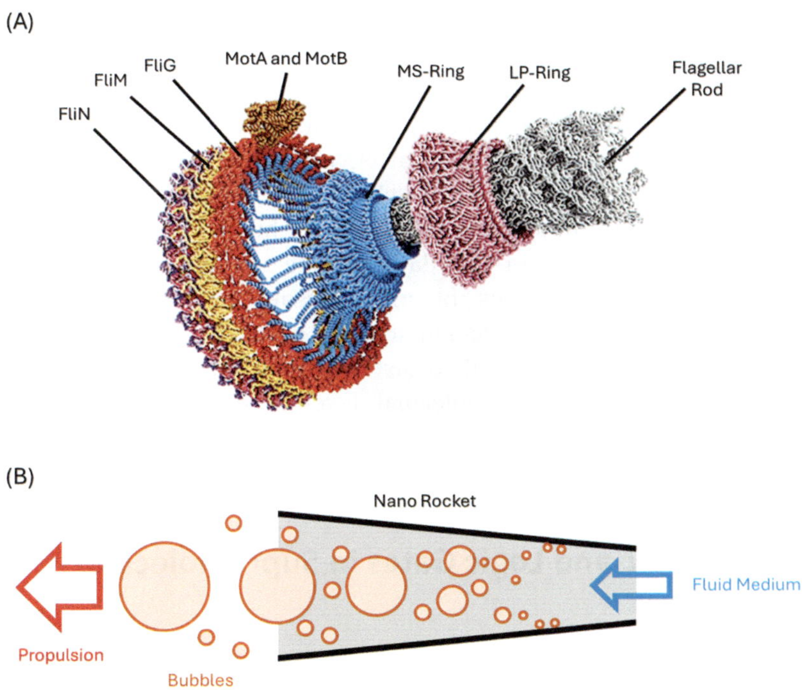

Figure 4.9 Examples of nanomachines: (A) A molecular motor of bacterial flagellar and (B) a nano rocket. Subplot (A) is remade with the permission under CC BY 4.0.[16]

MotA and MotB proteins form the stator MotAB complex of the bacterial flagellar motor.[16] This complex is embedded in the inner membrane and functions as the torque generator for the motor.[16] The proteins form channels that allow protons (H^+) or sodium ions (Na^+) to flow across the bacterial inner membrane, driven by an electrochemical gradient. This ion flow generates torque that interacts with the rotor's C-ring (FliG) to drive flagellar rotation. The LP-ring is a structural component located in the bacterial outer membrane and periplasmic space. It functions as a bushing or bearing that supports the rotation of the flagellar filament while maintaining its stability.[16]

Another interesting example is the nano rocket as shown in Fig. 4.9(B). This is a ramjet with no moving parts but can intake the fluid medium at front and generate gas (such as bubbles if the medium is liquid) to make propulsion at the tail.[13,17] Gas generation is based on the inner surface reacting with the medium. For example, $2Al + 6H_2O \Rightarrow 2Al(OH)_3 + 3H_2$ can make the H_2 bubble in water.[17]

Nanomachines have significant implications for biomedical engineering, including targeted drug delivery, where molecular machines can navigate biological environments to deliver drugs directly to diseased cells. In electronics, molecular switches and rotors can potentially be used to build components for quantum computers or for data storage systems that operate at unprecedented scales of miniaturization. Furthermore, in environmental science, nanomachines could be engineered to perform precise chemical transformations, such as breaking down pollutants at the molecular level or performing catalytic functions to synthesize chemicals in an environmentally friendly manner.

4.5 Logic and Logic Gates in Supramolecular CPN

In this section, we will discuss using supramolecules in intelligent plasmas. We will see that these molecules can provide direct and programmable logic operations in addition to the ML provided by the CPN.

4.5.1 The "if" Condition

First, let's consider a molecular switch,[12,14,18] such as a two-station rotaxanes, a.k.a. [2]rotaxanes, as shown in Fig. 4.10. The macrocycle molecule is "cyclobis(paraquat-p-phenylene)" or briefly "$CBPQT^{4+}$" molecule.[12,14,18] It is a cyclic molecule composed of four aromatic rings linked together by ethylene bridges, forming a square-like structure. $CBPQT^{4+}$ functions as a molecular switch primarily through its selective binding and unbinding to guest molecules based on changes in environmental conditions such as redox state, pH, or chemical stimuli. Tetrathiafulvalene (TTF) is an electron-rich molecule that can strongly interact with $CBPQT^{4+}$ through π-π stacking and charge-transfer interactions.[12,14,18] It acts as one of the "stations" where the $CBPQT^{4+}$ ring can reside. Similarly, 1,5-dioxynaphthalene (DNP) is another electron-rich aromatic system that can interact with $CBPQT^{4+}$ but generally less strongly than TTF.[12,14,18] It serves as the second "station" for the rotaxane's switching behavior. When TTF is oxidized, it becomes electron-deficient and loses its strong interaction with $CBPQT^{4+}$. This oxidation weakens the affinity between the TTF station and the $CBPQT^{4+}$ ring, prompting the ring to move to the DNP station. Oppositely, upon reduction back to its neutral state, TTF regains its electron-rich character, restoring its stronger interaction with $CBPQT^{4+}$, which can then move back to the TTF station. Therefore, such a switching mechanism allows the two-station rotaxane to act as a molecular switch, where the position of the $CBPQT^{4+}$ ring is controlled by the redox state of the TTF unit.

In plasma, oxidation can be achieved via electron impacts. High-energy electrons could potentially ionize the TTF, converting it to TTF^{+} or even TTF^{2+}. This oxidation would decrease the electron density around the TTF and thereby reduce its binding affinity with the $CBPQT^{4+}$, causing the ring to move to another station, such as DNP. The effectiveness of this process depends on the energy of the electrons in the plasma. For electron impact ionization to occur, the electron energy must exceed the ionization potential of the molecule. Typically, the ionization potential for organic molecules like TTF is in the range of 6–9 eV. Plasmas with electrons at energy higher than this can induce oxidation. Therefore, the redox state can be controlled by adjusting the plasma conditions (e.g., electron temperature or density) to regulate the rate of electron impact ionization. Similarly, ion collisions can also be used to control the redox status, if ions in the plasma are reactive enough. For instance: positive ions or radicals such as oxygen ions or hydroxyl radicals could oxidize the TTF component of the switch, while certain negative ions could reduce an oxidized species back to its neutral state. Therefore, collision energy and reactivity are crucial. Ions must have sufficient kinetic energy or chemical potential to affect the oxidation state of the molecular switch. Also, excited states, especially the metastable states with long lifetime, have enough internal energy to induce electron transfer processes that could alter the redox state of molecules in their vicinity. In addition, another mechanism to trigger the switch is UV emissions. UV light from plasma can excite or ionize molecules like TTF directly, causing oxidation. Conversely, UV-induced photochemistry could also lead to reduction if the right sensitizers or photo-redox catalysts are present. For example, the energy of 6.8–7.0 eV corresponding to UV light with a wavelength of approximately 177–182 nm can Oxidation of TTF to TTF^{+}. Excitation of TTF with 300–400 nm, corresponding to 3.1–4.1 eV, can also alter the binding affinity with the $CBPQT^{4+}$ on TTF.

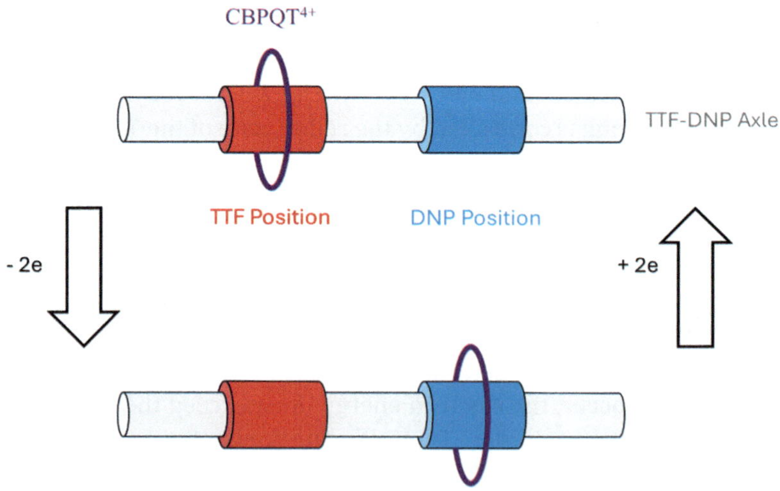

Figure 4.10 A two-station rotaxanes: The $CBPQT^{4+}$ and a TTF-DNP axle molecule.

This leads to an "if" condition addon to the CPN. For example, one can have a SiO_2 container connected to the [2] rotaxanes switch. The container is covered by the $CBPQT^{4+}$ ring. Once the switch is triggered, the container is opened, and the molecule stored in it will be released to be involved into a further program or CPN reaction paths.

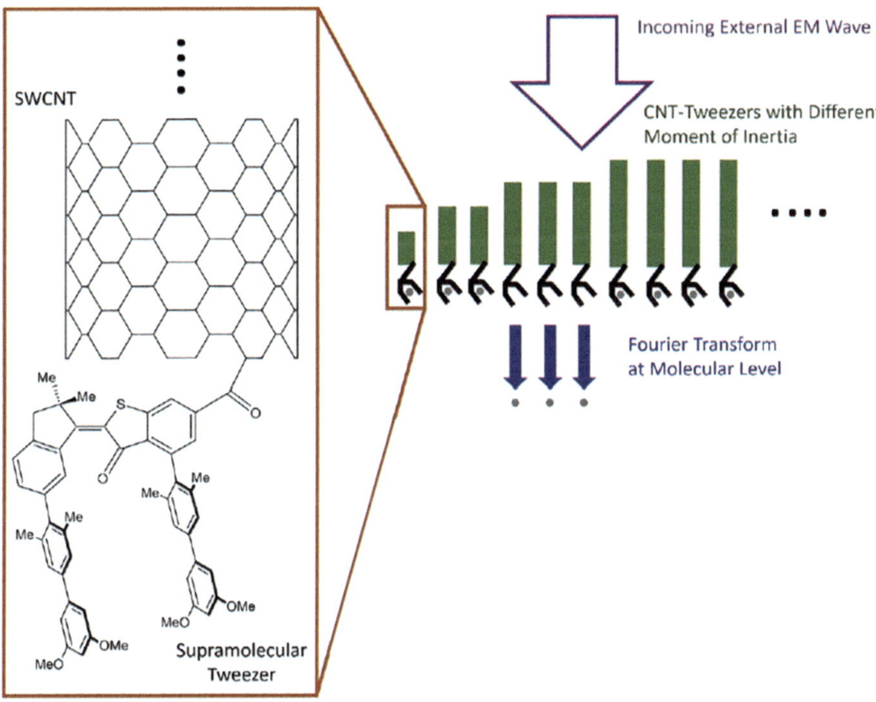

Figure 4.11 A molecular analog computer of discrete Fourier transform. The figure is remade with the permission under CC BY 4.0.[19]

A fantastic example of using the "if" condition can be a molecular analog computer of discrete Fourier transform[19] as shown in Fig. 4.11. The system is a set of supramolecules, each of them is a carbon nanotube (CNT) connected to a molecular squeeze holding a signal molecule.[19] Each of the supramolecule has a unique CNT length corresponding to an RF wavelength.[19] When an RF emission is incoming with multiple peaks in its spectrum, the corresponding CNTs will resonate, with enough energy to disturbance the molecular squeezes and release their signal molecules.[19] Therefore, each supramolecule here is an "if" condition switch. If the signal molecules are all identical, the resulting spatial distribution of them thus directly indicates the resulting spectrum, completing a discrete Fourier transform.[19] If the signal molecules are different corresponding to each length of CNT, the resulting concentration distribution over the signal molecule species space is thus indicating the resulting spectrum.[19] Please note that there are many different ways to achieve such an example; the connection between the CNT and the molecular squeeze does not have to be the same as shown in Fig. 4.11. One can even use other mechanisms of "if" condition to achieve such a molecular analog computer of Fourier transform.

4.5.2 The Logic Gates

In addition to the "if" condition, one may also be interested in making a "for" loop. However, there is no need to specifically design a "for" loop supramolecules because the CPN is a recurrent network, it is working naturally in a loop. We now will focus on how to build the logic gates based on the supramolecules-plasma combined system.

First, the most straightforward one is the OR gate, when two or multiple switches coexist in plasma. As shown in Fig. 4.12(A), take the [2]rotaxane example again: Two [2]rotaxane switches are connected with SiO_2 containers holding signal molecules A. Each container is covered by a $CBPQT^{4+}$ ring. When they are placed in plasma, any one of them is collided with by an ion will change the redox status and trigger the switch. The $CBPQT^{4+}$ ring will be moved away and release a molecule A as the result. The AND gate is slightly more complicated. Two [2]rotaxane switches are both connected with SiO_2 containers holding reactant molecule A and B respectively. The result of this system is the product C of a chemical reaction $A + B \Rightarrow C$. Therefore, only when both the two switches are triggered, the two $CBPQT^{4+}$ rings will be moved away and release both reactants to make the chemical reaction happen. This is an example of AND gate.

The above examples can be expanded to design the NOR gate and NAND gate. For a NOR gate, the result will be the existence of a molecule B, and in the CPN, we have the reaction A + B ⇒ C. In other words, the release of molecule A will consume the molecule B. As a result, any one of the [2]rotaxane switches triggered will lead to the loss of molecule B. This means a NOR gate. Similarly, for a NAND gate, we need to add another reaction C + D ⇒ E and the result is defined as the existence of the molecule D. When both the [2]rotaxane switches are triggered, we have A + B ⇒ C, and followed by the consumption of the molecule D, leading to a zero result. Of course, if we have only one of the switches triggered, we won't have the molecule C formed to consume D. This means a NAND gate.

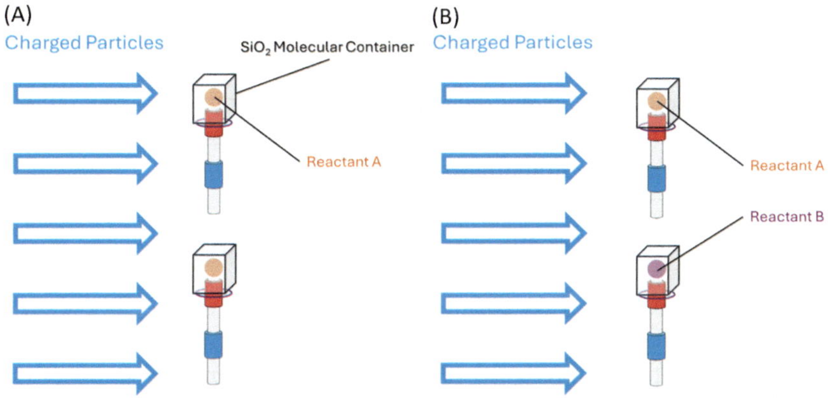

Figure 4.12 The examples of an OR gate: (A) Two [2]rotaxane switches are exposed in the plasma parallelly; and an AND gate: (B) Two [2]rotaxane switches with SiO_2 containers holding two reactant molecules.

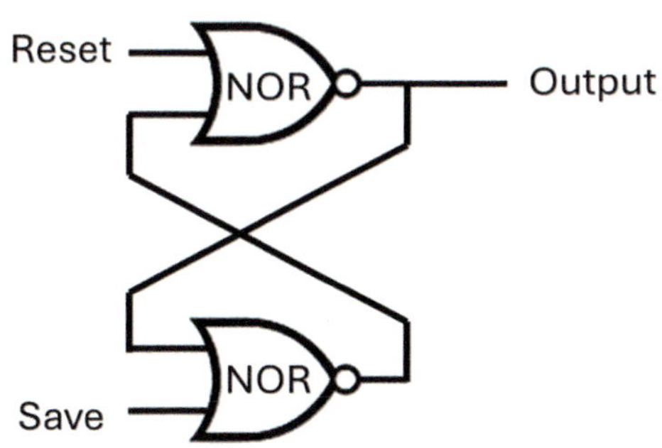

Figure 4.13 An SR latch that is a memory unit based on two NOR gates.

Based on the above gates, we can easily build XOR and XNOR like the regular circuit designs. Also, we can design a memory system based on these logic gates as shown in Fig. 4.13. The output will always be of the same value as the save side when the reset is zero, this is the data saving mechanism. When the reset is one, the output will always be zero, this is the data clearing mechanism.

4.6 Perspectives of the Intelligent Plasma and Material-Based AI

The theory of using "intelligent matter" to achieve artificial intelligence represents a groundbreaking shift in the future potential of technological development. By leveraging the natural chemical complexities and reactions within various materials, this approach could revolutionize how we think about computation, AI, and the fundamental building blocks of technology.

4.6.1 A New Paradigm in AI and Computing

The development of AI based on material substrates rather than digital semiconductors introduces a new paradigm in computing. This concept envisions a world where AI can be embedded directly into the materials themselves, leading to programmable matter that can perform complex tasks at a molecular level. Unlike traditional computing, which relies on semiconductor gates and binary logic, material-based AI would function through the intricate dance of chemical reactions and particle interactions. This could lead to AI systems that are not only more efficient but also capable of operating in environments where traditional computers fail, such as extreme temperatures, high radiation, or corrosive conditions.

Moreover, as semiconductor technology approaches its physical limits due to quantum uncertainties at the nanometer scale, intelligent matter offers a pathway to break free from the constraints of Moore's Law. Material-based computing does not rely on miniaturizing transistors but instead harnesses the natural, parallel-processing capabilities of chemical networks.

This could result in faster, more efficient processors that operate on principles fundamentally different from those of silicon chips, potentially leading to exponential growth in computational power without the heat and power constraints of current technology.

The concept of intelligent matter introduces the possibility of new states of matter, where materials possess inherent logical and strategic behaviors. This could fundamentally alter our understanding of physics and chemistry, merging them with concepts from computer science and AI. Such materials could lead to new forms of machines that do not require external programming but are capable of learning and evolving based on their chemical environment, opening possibilities for autonomous exploration and adaptation in unknown environments, such as deep space or the ocean floor. While the potential of intelligent matter is immense, there are significant challenges to overcome. Understanding and precisely controlling the chemical pathways and reactions that drive computation in intelligent matter will require advancements in chemistry, materials science, and machine learning. Additionally, ethical considerations will need to be addressed, particularly if intelligent materials start to exhibit behaviors that suggest a form of awareness or autonomy.

4.6.2 Applications Across Diverse Fields

The potential applications of intelligent matter span a wide range of fields. First, it could be used to develop smart drugs or implants that respond dynamically to changes in a patient's condition, allowing for real-time, personalized medical treatments. For example, a material embedded in the body could monitor biochemical signals and release medication as needed, or even perform diagnostic tasks autonomously. Also, in materials science and engineering, programmable matter could lead to self-assembling materials that change properties or functions based on environmental stimuli. This could revolutionize fields like aerospace, where materials are needed to adapt to different conditions during flight, or in consumer electronics, where devices could self-repair or transform according to user needs. This novel intelligent matter could lead to new forms of data storage and processing, where information is encoded in the chemical states or molecular arrangements of materials. This could result in highly compact, energy-efficient storage solutions or novel ways of transmitting data through chemical or physical changes rather than electronic signals. In addition, the material-based AI systems could be designed to be more sustainable than their digital counterparts. Since these systems do not require the same level of precision manufacturing and can operate in a broader range of conditions, they could be produced with lower energy inputs and fewer toxic byproducts. Additionally, the ability to operate in diverse environments might reduce the need for specialized infrastructure, further lowering the environmental footprint.

4.6.3 The Potential of Combining Supramolecule Logic with CPN

Combining supramolecular chemistry with material-based intelligence could significantly enhance the power and versatility of AI systems. As we introduced, supramolecules, which can perform logical operations and "if" conditions through their dynamic, non-covalent interactions, could act as molecular logic gates within a CPN. This integration allows for more sophisticated decision-making and computational processes directly within the material, effectively enabling it to execute complex, conditional operations at the molecular level. For example, a supramolecular-enhanced CPN could develop sophisticated pattern recognition skills, solve complex optimization problems, or even execute higher-level cognitive functions that involve multiple layers of reasoning and decision-making. This could revolutionize fields such as autonomous robotics, where materials need to adapt rapidly to changing environments, or smart materials that can autonomously optimize their properties for specific tasks. Such systems could process vast amounts of information simultaneously and adapt their computational pathways in real time based on environmental stimuli, surpassing the limitations of traditional silicon-based computers.

This fusion of supramolecular chemistry and CPN-based AI would lead to highly adaptive, multifunctional materials capable of learning and evolving far beyond current capabilities.

These intelligent materials could revolutionize various fields by performing diverse tasks without needing reprogramming, from autonomous environmental sensing to advanced machine learning applications. By embedding intelligence directly into the fabric of materials, we could create a new class of AI that operates faster and more efficiently, potentially leading to breakthroughs in robotics, medicine, and other advanced technologies, all while moving beyond the constraints of traditional digital computing.

References

1. Lin, L.; Keidar, M. Artificial Intelligence without Digital Computers: Programming Matter at a Molecular Scale. *Adv. Intell. Syst.* **2022**, *4* (11), 2200157. https://doi.org/10.1002/aisy.202200157.
2. Start, D. F. H.; Jacquinot, J.; Bergeaud, V.; Bhatnagar, V. P.; Cottrell, G. A.; Clement, S.; Eriksson, L.-G.; Fasoli, A.; Gondhalekar, A.; Gormezano, C.; Grosshoeg, G.; Guenther, K.; Harbour, P.; Horton, L. D.; Howman, A.; Jackel, H.; Jarvis, O. N.; Lawson, K. D.; Lowry, C.; Mantsinen, M.; Marcus, F. B.; Monk, R.; Righi, E.; Rimini, F. G.; Sadler, G. J.; Saibene, G. R.; Sartori, R.; Schunke, B.; Sharapov, S.; Sips, A. C. C.; Stamp, M.; Van Belle, P. D-T Fusion with Ion Cyclotron Resonance Heating in the JET Tokamak. *Phys. Rev. Lett.* **1998**, *80* (21), 4681–4684. https://doi.org/10.1103/PhysRevLett.80.4681.
3. Van Eester, D.; Lerche, E.; Ragona, R.; Messiaen, A.; Wauters, T.; JET contributors. Ion Cyclotron Resonance Heating Scenarios for DEMO. *Nucl. Fusion* **2019**, *59* (10), 106051. https://doi.org/10.1088/1741-4326/ab318b.
4. Pan, D.; Xu, H.; García de Abajo, J. Rotational Doppler Cooling and Heating. *Sci. Adv.* **2021**, *7*, eabd6705.
5. Tayi, A. S.; Kaeser, A.; Matsumoto, M.; Aida, T.; Stupp, S. I. Supramolecular Ferroelectrics. *Nat. Chem.* **2015**, *7* (4), 281–294. https://doi.org/10.1038/nchem.2206.
6. Fan, K.; Yang, J.; Wang, X.; Song, J. Rational Construction of Gel-Based Supramolecular Logic Gates by Using a Functional Gelator with Multiple-Stimuli Responsive Properties. *Soft Matter* **2014**, *10* (41), 8370–8375. https://doi.org/10.1039/C4SM01574F.
7. Leung, F. K.-C.; Ishiwari, F.; Kajitani, T.; Shoji, Y.; Hikima, T.; Takata, M.; Saeki, A.; Seki, S.; Yamada, Y. M. A.; Fukushima, T. Supramolecular Scaffold for Tailoring the Two-Dimensional Assembly of Functional

Molecular Units into Organic Thin Films. *J. Am. Chem. Soc.* **2016**, *138* (36), 11727–11733. https://doi.org/10.1021/jacs.6b05513.

8. Zinober, R. C.; Brockwell, D. J.; Beddard, G. S.; Blake, A. W.; Olmsted, P. D.; Radford, S. E.; Smith, D. A. Mechanically Unfolding Proteins: The Effect of Unfolding History and the Supramolecular Scaffold. *Protein Sci.* **2002**, *11* (12), 2759–2765. https://doi.org/10.1110/ps.0224602.
9. Magana, J. R.; Gumí-Audenis, B.; Tas, R. P.; Gascoigne, L.; Atkins, D. L.; Voets, I. K. Bioinspired Scaffolding by Supramolecular Amines Allows the Formation of One- and Two-Dimensional Silica Superstructures. *Chem. – Eur. J.* **2020**, *26* (66), 15330–15336. https://doi.org/10.1002/chem.202003139.
10. Amabilino, D. B.; Smith, D. K.; Steed, J. W. Supramolecular Materials. *Chem. Soc. Rev.* **2017**, *46* (9), 2404–2420. https://doi.org/10.1039/C7CS00163K.
11. Mattia, E.; Otto, S. Supramolecular Systems Chemistry. *Nat. Nanotechnol.* **2015**, *10* (2), 111–119. https://doi.org/10.1038/nnano.2014.337.
12. Nguyen, T. D.; Tseng, H.-R.; Celestre, P. C.; Flood, A. H.; Liu, Y.; Stoddart, J. F.; Zink, J. I. A Reversible Molecular Valve. *Proc. Natl. Acad. Sci.* **2005**, *102* (29), 10029–10034. https://doi.org/10.1073/pnas.0504109102.
13. Balzani, V.; Credi, A.; Raymo, F. M.; Stoddart, J. F. Artificial Molecular Machines. *Angew. Chem.* **2000**, *39* (19), 3348–3391. https://doi.org/10.1002/1521-3773(20001002)39:19<3348::AID-ANIE3348>3.0.CO;2-X.
14. Gao, W.; Wang, J. The Environmental Impact of Micro/Nanomachines: A Review. *ACS Nano* **2014**, *8* (4), 3170–3180. https://doi.org/10.1021/nn500077a.
15. Feynman, R. There's Plenty of Room at the Bottom. *Reson. J. Sci. Educ.* **2011**, *16* (9).
16. Singh, P. K.; Sharma, P.; Afanzar, O.; Goldfarb, M. H.; Maklashina, E.; Eisenbach, M.; Cecchini, G.; Iverson, T. M. CryoEM Structures Reveal How the Bacterial Flagellum Rotates and Switches Direction. *Nat. Microbiol.* **2024**, *9* (5), 1271–1281. https://doi.org/10.1038/s41564-024-01674-1.
17. Li, J.; Rozen, I.; Wang, J. Rocket Science at the Nanoscale. *ACS Nano* **2016**, *10* (6), 5619–5634. https://doi.org/10.1021/acsnano.6b02518.

18. Wang, J. *Nanomachines: Fundamentals and Applications*; John Wiley & Sons, 2013.
19. Lin, L.; Keidar, M. An Intelligent Material with Chemical Pathway Networks. arXiv preprint arXiv:2109.02735, **2021**.

Appendix: A Kinetic Scheme of Helium–Air Plasma

Li Lin

School of Engineering and Applied Science, The George Washington University, Washington, DC 20052, USA

lilin@email.gwu.edu

Category	Reaction Number	Collision	Rate Coefficient	Refs
e–He	1	$e + He^{+} \Rightarrow He$	$2.7 \times 10^{-13} (T_e k_B/e)^{-3/4}$	[1]
	2	$2e + He^{+} \Rightarrow He\ (2^1P) + e$	$1.1 \times 10^{-14} T_e^{-4.5} T_g^{2.3}$	[1]
	3	$2e + He^{+} \Rightarrow He\ (2^3P) + e$	$1.1 \times 10^{-14} T_e^{-4.5} T_g^{2.3}$	[1]
	4	$2e + He^{+} \Rightarrow He\ (2^1S) + e$	$1.1 \times 10^{-14} T_e^{-4.5} T_g^{2.3}$	[1]
	5	$2e + He^{+} \Rightarrow He\ (2^3S) + e$	$1.1 \times 10^{-14} T_e^{-4.5} T_g^{2.3}$	[1]
	6	$e + He \Rightarrow He\ (2^3S) + e$	BOLSIG+	[2]
	7	$e + He \Rightarrow He\ (2^1S) + e$	BOLSIG+	[2]
	8	$e + He \Rightarrow He\ (2^3P) + e$	BOLSIG+	[2]
	9	$e + He \Rightarrow He\ (2^1P) + e$	BOLSIG+	[2]
	10	$e + He \Rightarrow He\ (3^3S) + e$	BOLSIG+	[2]
	11	$e + He \Rightarrow He\ (3^1S) + e$	BOLSIG+	[2]
	12	$e + He \Rightarrow He\ (3^3P) + e$	BOLSIG+	[2]
	13	$e + He \Rightarrow He\ (3^3D) + e$	BOLSIG+	[2]
	14	$e + He \Rightarrow He\ (3^1P) + e$	BOLSIG+	[2]
	15	$e + He \Rightarrow He\ (3^1D) + e$	BOLSIG+	[2]

Machine Learning Plasmas and the Neuromorphic Plasma Chemistry
Edited by Li Lin and Michael Keidar

ISBN 978-981-5129-84-7 (Hardcover), 978-1-003-71013-4 (eBook)
www.jennystanford.com

Category	Reaction Number	Collision	Rate Coefficient	Refs
	16	e + He ⇒ He (4^3S) + e	BOLSIG+	[2]
	17	e + He ⇒ He (4^1S) + e	BOLSIG+	[2]
	18	e + He ⇒ He (4^3P) + e	BOLSIG+	[2]
	19	e + He ⇒ He (4^1D) + e	BOLSIG+	[2]
	20	e + He ⇒ He (4^3D) + e	BOLSIG+	[2]
	21	e + He ⇒ He (4^1F) + e	BOLSIG+	[2]
	22	e + He ⇒ He (4^3F) + e	BOLSIG+	[2]
	23	e + He ⇒ He (4^1P) + e	BOLSIG+	[2]
	24	e + He ⇒ He^+ + 2e	BOLSIG+	[2]
	25	e + He (2^1P) ⇒ He^+ + 2e	$1.28 \times 10^{-7} (T_e k_B/e)^{0.6} \exp(-4.78/ (T_e k_B/e))$	[1]
	26	e + He (2^3P) ⇒ He^+ + 2e	$1.28 \times 10^{-7} (T_e k_B/e)^{0.6} \exp(-4.78/ (T_e k_B/e))$	[1]
	27	e + He (2^1S) ⇒ He^+ + 2e	$1.28 \times 10^{-7} (T_e k_B/e)^{0.6} \exp(-4.78/ (T_e k_B/e))$	[1]
	28	e + He (2^3S) ⇒ He^+ + 2e	$1.28 \times 10^{-7} (T_e k_B/e)^{0.6} \exp(-4.78/ (T_e k_B/e))$	[1]
	29	e + He (2^1P) ⇒ He (2^3P) + e	$1.8 \times 10^{-7} (1+0.47 (T_e k_B/e))^{-1}$	[1]
	30	e + He (2^1P) ⇒ He (2^1S) + e	$3.7 \times 10^{-7} (1+0.37 (T_e k_B/e))^{-1}$	[1]
	31	e + He (2^1P) ⇒ He (2^3S) + e	$3.1 \times 10^{-8} (1+0.13 (T_e k_B/e)^2)^{-1}$	[1]
	32	e + He (2^1P) ⇒ He + e	5×10^{-10}	[1]
	33	e + He (2^3P) ⇒ He (2^1S) + e	$3.1 \times 10^{-8} (1+0.94 (T_e k_B/e))^{-1}$	[1]
	34	e + He (2^3P) ⇒ He (2^3S) + e	$2.2 \times 10^{-7} (T_e k_B/e)^{1/3}$	[1]
	35	e + He (2^3P) ⇒ He + e	2.4×10^{-10}	[1]
	36	e + He (2^1S) ⇒ He (2^3S) + e	$3.8 \times 10^{-7} (1+T_e k_B/e)^{-1}$	[1]
	37	e + He (2^1S) ⇒ He + e	3×10^{-9}	[1]
	38	e + He (2^3S) ⇒ He + e	$4.1 \times 10^{-9} (1+T_e k_B/e)^{-1}$	[1]
	39	e + He_2^+ ⇒ He (2^1P) + He	$8.9 \times 10^{-9} (T_g/T_e)^{1.5}$	[1]
	40	e + He_2^+ ⇒ He (2^3P) + He	$8.9 \times 10^{-9} (T_g/T_e)^{1.5}$	[1]

Category	Reaction Number	Collision	Rate Coefficient	Refs
	41	$e + He_2^+ \Rightarrow He\ (2^1S) + He$	$8.9 \times 10^{-9}\ (T_g/T_e)^{1.5}$	[1]
	42	$e + He_2^+ \Rightarrow He\ (2^3S) + He$	$8.9 \times 10^{-9}\ (T_g/T_e)^{1.5}$	[1]
	43	$e + He + He_2^+ \Rightarrow He\ (2^1P) + 2He$	$10^{-26}\ (T_g/T_e)^2$	[1]
	44	$e + He + He_2^+ \Rightarrow He\ (2^3P) + 2He$	$10^{-26}\ (T_g/T_e)^2$	[1]
	45	$e + He + He_2^+ \Rightarrow He\ (2^3S) + 2He$	$10^{-26}\ (T_g/T_e)^2$	[1]
e–H	46	$2e + H^+ \Rightarrow H + e$	$10^{-19}\ (T_g/T_e)^{4.5}$	[3]
e–H_2	47	$e + H_2^+ \Rightarrow 2H$	$1.86 \times 10^{-7}/T_e^{0.43}$	[3]
	48	$2e + H_2^+ \Rightarrow H_2 + e$	$10^{-19}\ (T_g/T_e)^{4.5}$	[3]
	49	$e + H_2 \Rightarrow 2H + e$	$3.29 \times 10^{-9}(1.5T_ek_B/e)^{0.578}$ $\exp(-7.56/(1.5T_ek_B/e))$	[3]
	50	$e + H_2 \Rightarrow H_2^+ + 2e$	$4 \times 10^{-11}\ (1.5T_ek_B/e)^{2.13}$ $\exp(-14.9/(1.5T_ek_B/e))$	[3]
e–C	51	$e + C^+ \Rightarrow C$	4.51×10^{-12}	[4]
	52	$e + C \Rightarrow C\ (2s2p^2\ ^1D) + e$	BOLSIG+	[2]
	53	$e + C \Rightarrow C\ (2s2p^2\ ^1S) + e$	BOLSIG+	[2]
	54	$e + C \Rightarrow C\ (2s2p^3\ ^5S^o) + e$	BOLSIG+	[2]
	55	$e + C \Rightarrow C\ (2s2p3s\ ^3P^o) + e$	BOLSIG+	[2]
	56	$e + C \Rightarrow C\ (2s2p3s\ ^1P^o) + e$	BOLSIG+	[2]
	57	$e + C \Rightarrow C\ (2s2p^3\ ^3D^o) + e$	BOLSIG+	[2]
	58	$e + C \Rightarrow C\ (2s2p3p\ ^1P) + e$	BOLSIG+	[2]
	59	$e + C \Rightarrow C\ (2s2p3p\ ^3D) + e$	BOLSIG+	[2]
	60	$e + C \Rightarrow C\ (2s2p3p\ ^3S) + e$	BOLSIG+	[2]
	61	$e + C \Rightarrow C\ (2s2p3p\ ^3P) + e$	BOLSIG+	[2]
	62	$e + C \Rightarrow C\ (2s2p3p\ ^1D) + e$	BOLSIG+	[2]
	63	$e + C \Rightarrow C\ (2s2p3p\ ^1S) + e$	BOLSIG+	[2]
	64	$e + C \Rightarrow C\ (2s2p^3\ ^3P^o) + e$	BOLSIG+	[2]
	65	$e + C \Rightarrow C\ (2s2p3d\ ^1D^o) + e$	BOLSIG+	[2]
	66	$e + C \Rightarrow C\ (2s2p4s\ ^3P^o) + e$	BOLSIG+	[2]

Category	Reaction Number	Collision	Rate Coefficient	Refs
	67	$e + C \Rightarrow C\ (2s2p3d\ ^3F^o) + e$	BOLSIG+	[2]
	68	$e + C \Rightarrow C\ (2s2p3d\ ^3D^o) + e$	BOLSIG+	[2]
	69	$e + C \Rightarrow C\ (2s2p4s\ ^1P^o) + e$	BOLSIG+	[2]
	70	$e + C \Rightarrow C\ (2s2p3d\ ^1F^o) + e$	BOLSIG+	[2]
	71	$e + C \Rightarrow C\ (2s2p3d\ ^1P^o) + e$	BOLSIG+	[2]
	72	$e + C \Rightarrow C\ (2s2p3d\ ^3P^o) + e$	BOLSIG+	[2]
	73	$e + C \Rightarrow C^+ + 2e$	BOLSIG+	[2]
e–N	74	$e + N^+ + M \Rightarrow N + M$	$3.12 \times 10^{-23}/T_e^{1.5}$	[3]
	75	$2e + N^+ \Rightarrow N + e$	$10^{-19}\ (T_g/T_e)^{4.5}$	[3]
	76	$e + N \Rightarrow N^+ + 2e$	$1.45 \times 10^{-11} \times (1.5T_e k_B/e)^{2.58} \exp(-8.54/(1.5T_e k_B/e))$	[3]
e–N_2	77	$e + N_2^+ \Rightarrow 2N$	$2.8 \times 10^{-7}\ (300/T_e)^{0.5}$	[5]
	78	$e + N_2^+ \Rightarrow N_2^+\ (B^2\Sigma_u^+ v0) + e$	$0.8765807 + 2.8182/(1 + (T_e k_B/e/5.78903)^{2.997126}) \times 100 \times (2T_e k_B/me)^{0.5} \times 10^{-6}$	[6]*
	79	$e + N_2^+ + M \Rightarrow N_2 + M$	$3.12 \times 10^{-23}/T_e^{1.5}$	[3]
	80	$2e + N_2^+ \Rightarrow N_2 + e$	$10^{-19}\ (T_g/T_e)^{4.5}$	[3]
	81	$e + N_2 \Rightarrow N_2^+ + 2e$	BOLSIG+	[2]
	82	$e + N_2 \Rightarrow N_2\ (B^3\Pi_g v0) + e$	BOLSIG+	[2]
	83	$e + N_2 \Rightarrow N_2\ (B^3\Pi_g v1) + e$	BOLSIG+	[2]
	84	$e + N_2 \Rightarrow N_2\ (B^3\Pi_g v2) + e$	BOLSIG+	[2]
	85	$e + N_2 \Rightarrow N_2\ (B^3\Pi_g v3) + e$	BOLSIG+	[2]
	86	$e + N_2 \Rightarrow N_2\ (B^3\Pi_g v4) + e$	BOLSIG+	[2]
	87	$e + N_2 \Rightarrow N_2\ (C^3\Pi_u v0) + e$	BOLSIG+	[2]
	88	$e + N_2 \Rightarrow N_2\ (C^3\Pi_u v1) + e$	BOLSIG+	[2]
	89	$e + N_2 \Rightarrow N_2\ (C^3\Pi_u v2) + e$	BOLSIG+	[2]
e–CO	90	$e + CO^+ \Rightarrow C + O$	$\max [1.7778 \times 10^{-5} (T_e k_B/e) - 2.5556 \times 10^{-6}, 0]$	[7]
	91	$e + CO \Rightarrow C + O^-$	BOLSIG+	[2]

Category	Reaction Number	Collision	Rate Coefficient	Refs
	92	$e + CO \Rightarrow CO\ (A^1\Pi) + e$	BOLSIG+	[2]
	93	$e + CO \Rightarrow CO\ (B^1\Sigma^+) + e$	BOLSIG+	[2]
	94	$e + CO \Rightarrow C + O + e$	BOLSIG+	[2]
	95	$e + CO \Rightarrow CO\ (C^1\Sigma^+) + e$	BOLSIG+	[2]
	96	$e + CO \Rightarrow CO\ (E^1\Pi) + e$	BOLSIG+	[2]
	97	$e + CO \Rightarrow CO^+ + 2e$	BOLSIG+	[2]
	98	$e + CO \Rightarrow C^+ + O + 2e$	BOLSIG+	[2]
	99	$e + CO \Rightarrow C + O^+ + 2e$	BOLSIG+	[2]
e–O	100	$e + O \Rightarrow O + e + h\nu$ (777.4 nm)	$10^7 \exp(-75.42 - 1.44 (T_e k_B/e) \ln (T_e k_B/e) + 9 (T_e k_B/e)^{0.5} \ln (T_e k_B/e)) \times 100 \times (2T_e k_B/m_e)^{0.5}$	[8]
	101	$e + O^+ + M \Rightarrow O + M$	$3.12 \times 10^{-23}/T_e^{1.5}$	[3]
	102	$2e + O^+ \Rightarrow O + e$	$10^{-19} (T_g/T_e)^{4.5}$	[3]
	103	$e + O \Rightarrow O^+ + 2e$	$4.75 \times 10^{-9} (1.5T_e k_B/e)^{0.61} \exp(-22.1/(1.5T_e k_B/e))$	[3]
	104	$e + O + O_2 \Rightarrow O^- + O_2$	10^{-31}	[3]
	105	$e + O + O_2 \Rightarrow O + O_2^-$	10^{-31}	[3]
e–O_2	106	$e + O_2^+ \Rightarrow 2O$	$2 \times 10^{-7} \times (300/T_e)$	[5]
	107	$2e + O_2^+ \Rightarrow O_2 + e$	$10^{-19} (T_g/T_e)^{4.5}$	[3]
	108	$e + O_2^+ + M \Rightarrow O_2 + M$	$3.12 \times 10^{-23}/T_e^{1.5}$	[3]
	109	$e + O_2 \Rightarrow O^- + O$	BOLSIG+	[2]
	110	$e + O_2 \Rightarrow O_2^+ + 2e$	BOLSIG+	[2]
	111	$e + 2O_2 \Rightarrow O_2^- + O_2$	$1.4 \times 10^{-29} T_g/T_e \exp(-600/T_g) \exp(700 (T_e - T_g)/(T_e T_g))$	[3]
	112	$e + O_2 + N_2 \Rightarrow O_2^- + N_2$	$1.4 \times 10^{-31} (T_g/T_e)^2 \exp(-70/T_g) \exp(1500 (T_e - T_g)/(T_e T_g))$	[3]
e–CO_2	113	$e + CO_2 \Rightarrow CO + O^-$	BOLSIG+	[2]
	114	$e + CO_2 \Rightarrow CO_2^+ + 2e$	BOLSIG+	[2]
	115	$e + CO_2^+ \Rightarrow CO\ (A^1\Pi) + O$	4×10^{-7}	[9]
e–O_3	116	$e + O_3 \Rightarrow O_2^- + O$	10^{-9}	[5]
	117	$e + O_3 \Rightarrow O^- + O_2$	10^{-11}	[3]

Category	Reaction Number	Collision	Rate Coefficient	Refs
	118	$e + O_3 \Rightarrow O + O_2 + e$	$1.78 \times 10^{-6} (1.5T_e k_B/e)^{-0.614} \exp(-11.5/(1.5T_e k_B/e))$	[3]
	119	$e + O_3 + M \Rightarrow O_3^- + M$	10^{-31}	[3]
e–H_2O	120	$e + H_2O^+ \Rightarrow O + 2H$	$1.37 \times 10^{-6}/T_e^{0.5}$	[3]
	121	$e + H_2O^+ \Rightarrow O + H_2$	$1.37 \times 10^{-6}/T_e^{0.5}$	[3]
	122	$e + H_2O^+ \Rightarrow OH + H$	$2.73 \times 10^{-6}/T_e^{0.5}$	[3]
	123	$2e + H_2O^+ \Rightarrow H_2O + e$	$10^{-19} (T_g/T_e)^{4.5}$	[3]
	124	$e + H_2O \Rightarrow H_2O^+ + 2e$	$9.65 \times 10^{-18} (1.5T_e k_B/e)^{2.53} \exp(-8.99/(1.5T_e k_B/e))$	[3]
	125	$e + H_2O \Rightarrow H_2 + O^-$	$2.97 \times 10^{-15} (1.5T_e k_B/e)^{-1.56} \exp(-13.67/(1.5T_e k_B/e))$	[3]
	126	$e + H_2O \Rightarrow OH + H^-$	$4.42 \times 10^{-14} (1.5T_e k_B/e)^{-2.0} \exp(-13.39/(1.5T_e k_B/e))$	[3]
	127	$e + H_2O \Rightarrow OH^+ + H + 2e$	$9.89 \times 10^{-6} (1.5T_e k_B/e)^{-1.64} \exp(-67.6/(1.5T_e k_B/e))$	[3]
	128	$e + H_2O \Rightarrow H^+ + OH + 2e$	$7.45 \times 10^{-9} (1.5T_e k_B/e)^{0.34} \exp(-54.2/(1.5T_e k_B/e))$	[3]
	129	$e + H_2O \Rightarrow O^+ + H_2 + 2e$	$7.4 \times 10^{-10} (1.5T_e k_B/e)^{0.45} \exp(-55.5/(1.5T_e k_B/e))$	[3]
	130	$e + H_2O \Rightarrow H_2^+ + O + 2e$	$8.49 \times 10^{-9} (1.5T_e k_B/e)^{-1.23} \exp(-74/(1.5T_e k_B/e))$	[3]
	131	$e + H_2O \Rightarrow OH + H + e$	$5.15 \times 10^{-9} (1.5T_e k_B/e)^{0.62} \exp(-10.9/(1.5T_e k_B/e))$	[3]
	132	$e + H_2O \Rightarrow OH^- + H$	$9.6 \times 10^{-10} (1.5T_e k_B/e)^{-1.7} \exp(-13.31/(1.5T_e k_B/e))$	[3]
e–OH	133	$e + OH \Rightarrow OH\ (A^2\Sigma^+ v0) + e$	$1.17 \times 10^{-8} T_e^{0.52} \exp(-44642.8/T_e)$	[10]
	134	$2e + OH^+ \Rightarrow OH + e$	$10^{-19} (T_g/T_e)^{4.5}$	[3]
e–NO	135	$e + NO^+ \Rightarrow N + O$	$1.07 \times 10^{-5}/T_e^{0.85}$	[3]

Category	Reaction Number	Collision	Rate Coefficient	Refs
	136	$e + NO^+ + M \Rightarrow NO + M$	$3.12 \times 10^{-23}/T_e^{1.5}$	[3]
	137	$2e + NO^+ \Rightarrow NO + e$	$10^{-19} (T_g/T_e)^{4.5}$	[3]
	138	$e + NO + M \Rightarrow NO^- + M$	8×10^{-31}	[3]
e–NO_2	139	$e + NO_2^+ \Rightarrow NO + O$	$3.46 \times 10^{-6}/T_e^{0.5}$	[3]
	140	$e + NO_2 + M \Rightarrow NO_2^- + M$	1.5×10^{-30}	[3]
	141	$e + NO_2 \Rightarrow O^- + NO$	10^{-11}	[3]
e–N_2O	142	$e + N_2O^+ \Rightarrow N_2 + O$	$3.46 \times 10^{-6}/T_e^{0.5}$	[3]
	143	$e + N_2O \Rightarrow O^- + N_2$	2×10^{-10}	[3]
e–HNO_3	144	$e + HNO_3 \Rightarrow NO_2^- + OH$	5×10^{-8}	[3]
H^+	145	$H^+ + H^- \Rightarrow 2H$	$2 \times 10^{-7} (300/T_g)^{0.5}$	[3]
	146	$H^+ + H_2 \Rightarrow H + H_2^+$	6.4×10^{-10}	[3]
	147	$H^+ + O \Rightarrow O^+ + H$	3.8×10^{-10}	[3]
	148	$H^+ + O^- \Rightarrow O + H$	$2 \times 10^{-7} (300/T_g)^{0.5}$	[3]
	149	$H^+ + OH^- \Rightarrow OH + H$	$2 \times 10^{-7} (300/T_g)^{0.5}$	[3]
	150	$H^+ + H_2O \Rightarrow H_2O^+ + H$	8.2×10^{-9}	[3]
	151	$H^+ + NO \Rightarrow NO^+ + H$	1.9×10^{-9}	[3]
	152	$H^+ + NO^- \Rightarrow NO + H$	$2 \times 10^{-7} (300/T_g)^{0.5}$	[3]
	153	$H^+ + O_2 \Rightarrow O_2^+ + H$	1.17×10^{-9}	[3]
	154	$H^+ + O_2^- \Rightarrow O_2 + H$	$2 \times 10^{-7} (300/T_g)^{0.5}$	[3]
	155	$H^+ + N_2O^- \Rightarrow N_2O + H$	$2 \times 10^{-7} (300/T_g)^{0.5}$	[3]
	156	$H^+ + NO_2^- \Rightarrow NO_2 + H$	$2 \times 10^{-7} (300/T_g)^{0.5}$	[3]
	157	$H^+ + O_3^- \Rightarrow O_3 + H$	$2 \times 10^{-7} (300/T_g)^{0.5}$	[3]
	158	$H^+ + NO_3^- \Rightarrow NO_3 + H$	$2 \times 10^{-7} (300/T_g)^{0.5}$	[3]
H	159	$H + H^- \Rightarrow H_2 + e$	1.8×10^{-9}	[3]
	160	$H + O^+ \Rightarrow H^+ + O$	6.8×10^{-10}	[3]
	161	$H + O + M \Rightarrow OH + M$	1.62×10^{-32}	[3]
	162	$H + OH + M \Rightarrow H_2O + M$	$6.1 \times 10^{-26}/T_g^2$	[3]
	163	$H + OH^- \Rightarrow H_2O + e$	1.8×10^{-9}	[3]
	164	$H + NO + M \Rightarrow HNO + M$	$10^{-32} \exp(300/T_g)$	[3]
	165	$H + O_2 + M \Rightarrow HO_2 + M$	$5.4 \times 10^{-32} (T_g/300)^{-1.8}$	[3]
	166	$H + O_2^- \Rightarrow HO_2 + e$	1.4×10^{-9}	[3]

Category	Reaction Number	Collision	Rate Coefficient	Refs
	167	$2H + M \Rightarrow H_2 + M$	$1.8 \times 10^{-30}/T_g$	[3]
	168	$H + HNO \Rightarrow NO + H_2$	$3 \times 10^{-11} \exp(-500/T_g)$	[3]
	169	$H + HO_2 \Rightarrow H_2 + O_2$	5.6×10^{-12}	[3]
	170	$H + HO_2 \Rightarrow O + H_2O$	2.4×10^{-12}	[3]
	171	$H + HO_2 \Rightarrow 2OH$	$4.2 \times 10^{-10} \exp(-950/T_g)$	[3]
	172	$H + H_2O_2 \Rightarrow OH + H_2O$	$1.69 \times 10^{-11} \exp(-1800/T_g)$	[3]
	173	$H + H_2O_2 \Rightarrow HO_2 + H_2$	$2.8 \times 10^{-12} \exp(-1900/T_g)$	[3]
	174	$H + CO_2 \Rightarrow CO + OH$	2.51×10^{-10} $\exp(-13351/T_g)$	NIST
	175	$H + NO_2 \Rightarrow OH + NO$	1.47×10^{-10}	[3]
	176	$H + NO_2^- \Rightarrow OH^- + NO$	4×10^{-10}	[3]
	177	$H + HNO_2 \Rightarrow NO_2 + H_2$	$2 \times 10^{-11} \exp(-3700/T_g)$	[3]
	178	$H + O_3 \Rightarrow OH + O_2$	$2.8 \times 10^{-11} (T_g/300)^{0.75}$	[3]
	179	$H + O_3^- \Rightarrow OH^- + O_2$	8.4×10^{-10}	[3]
	180	$H + NO_3 \Rightarrow OH + NO_2$	$5.8 \times 10^{-10} \exp(-750/T_g)$	[3]
	181	$H + HNO_3 \Rightarrow NO_2 + H_2O$	$1.39 \times 10^{-14} (T_g/298)^{3.29}$ $\exp(-3160/T_g)$	[3]
H^-	182	$H^- + H_2^+ \Rightarrow H + H_2$	$2 \times 10^{-7} (300/T_g)^{0.5}$	[3]
	183	$H^- + H_2^+ \Rightarrow 3H$	10^{-7}	[3]
	184	$H^- + N^+ \Rightarrow N + H$	$2 \times 10^{-7} (300/T_g)^{0.5}$	[3]
	185	$H^- + O^+ \Rightarrow O + H$	$2 \times 10^{-7} (300/T_g)^{0.5}$	[3]
	186	$H^- + OH^+ \Rightarrow H + OH$	$2 \times 10^{-7} (300/T_g)^{0.5}$	[3]
	187	$H^- + OH^+ \Rightarrow 2H + O$	10^{-7}	[3]
	188	$H^- + NO^+ \Rightarrow H + NO$	$2 \times 10^{-7} (300/T_g)^{0.5}$	[3]
	189	$H^- + NO^+ \Rightarrow H + N + O$	10^{-7}	[3]
	190	$H^- + H_2O \Rightarrow OH^- + H_2$	3.8×10^{-9}	[3]
	191	$H^- + H_2O^+ \Rightarrow H + H_2O$	$2 \times 10^{-7} (300/T_g)^{0.5}$	[3]
	192	$H^- + H_2O^+ \Rightarrow 2H + OH$	10^{-7}	[3]
	193	$H^- + N_2^+ \Rightarrow H + N_2$	$2 \times 10^{-7} (300/T_g)^{0.5}$	[3]
	194	$H^- + N_2^+ \Rightarrow H + 2N$	10^{-7}	[3]
	195	$H^- + O_2^+ \Rightarrow H + O_2$	$2 \times 10^{-7} (300/T_g)^{0.5}$	[3]
	196	$H^- + O_2^+ \Rightarrow H + 2O$	10^{-7}	[3]

Category	Reaction Number	Collision	Rate Coefficient	Refs
	197	$H^- + N_2O^+ \Rightarrow H + N_2O$	$2 \times 10^{-7} (300/T_g)^{0.5}$	[3]
	198	$H^- + N_2O^+ \Rightarrow H + N_2 + O$	10^{-7}	[3]
	199	$H^- + N_2O \Rightarrow OH^- + N_2$	1.1×10^{-9}	[3]
	200	$H^- + NO_2^+ \Rightarrow H + NO_2$	$2 \times 10^{-7} (300/T_g)^{0.5}$	[3]
	201	$H^- + NO_2^+ \Rightarrow H + N + O_2$	10^{-7}	[3]
	202	$H^- + NO_2 \Rightarrow NO_2^- + H$	2.9×10^{-9}	[3]
H_2^+	203	$H_2^+ + O^- \Rightarrow O + H_2$	$2 \times 10^{-7} (300/T_g)^{0.5}$	[3]
	204	$H_2^+ + O^- \Rightarrow O + 2H$	10^{-7}	[3]
	205	$H_2^+ + OH^- \Rightarrow OH + H_2$	$2 \times 10^{-7} (300/T_g)^{0.5}$	[3]
	206	$H_2^+ + OH^- \Rightarrow OH + 2H$	10^{-7}	[3]
	207	$H_2^+ + H_2O \Rightarrow H_2O^+ + H_2$	3.86×10^{-9}	[3]
	208	$H_2^+ + NO^- \Rightarrow NO + H_2$	$2 \times 10^{-7} (300/T_g)^{0.5}$	[3]
	209	$H_2^+ + NO^- \Rightarrow NO + 2H$	10^{-7}	[3]
	210	$H_2^+ + O_2 \Rightarrow O_2^+ + H_2$	7.83×10^{-10}	[3]
	211	$H_2^+ + O_2^- \Rightarrow O_2 + H_2$	$2 \times 10^{-7} (300/T_g)^{0.5}$	[3]
	212	$H_2^+ + O_2^- \Rightarrow O_2 + 2H$	10^{-7}	[3]
	213	$H_2^+ + N_2O^- \Rightarrow N_2O + H_2$	$2 \times 10^{-7} (300/T_g)^{0.5}$	[3]
	214	$H_2^+ + N_2O^- \Rightarrow N_2O + 2H$	10^{-7}	[3]
	215	$H_2^+ + NO_2^- \Rightarrow NO_2 + H_2$	$2 \times 10^{-7} (300/T_g)^{0.5}$	[3]
	216	$H_2^+ + NO_2^- \Rightarrow NO_2 + 2H$	10^{-7}	[3]
	217	$H_2^+ + O_3^- \Rightarrow O_3 + H_2$	$2 \times 10^{-7} (300/T_g)^{0.5}$	[3]
	218	$H_2^+ + O_3^- \Rightarrow O_3 + 2H$	10^{-7}	[3]
	219	$H_2^+ + NO_3^- \Rightarrow NO_3 + H_2$	$2 \times 10^{-7} (300/T_g)^{0.5}$	[3]
	220	$H_2^+ + NO_3^- \Rightarrow NO_3 + 2H$	10^{-7}	[3]
H_2	221	$H_2 + O^+ \Rightarrow OH^+ + H$	1.7×10^{-9}	[3]
	222	$H_2 + O^- \Rightarrow OH^- + H$	3.3×10^{-11}	[3]
	223	$H_2 + O^- \Rightarrow H_2O + e$	7×10^{-10}	[3]
	224	$H_2 + OH^+ \Rightarrow H_2O^+ + H$	9.7×10^{-10}	[3]
	225	$H_2 + OH \Rightarrow H + H_2O$	$3.2 \times 10^{-11} \exp(-2600/T_g)$	[3]
He	226	$He + He (3^1S) \Rightarrow He_2^+ + e$	$3.6 \times 10^{-12} \times 10^{-2}$	[1] ‡
	227	$He + He (3^1P) \Rightarrow He_2^+ + e$	$1.11 \times 10^{-10} \times 10^{-2}$	[1] ‡
	228	$He + He (3^1D) \Rightarrow He_2^+ + e$	$7.36 \times 10^{-10} \times 10^{-2}$	[1] ‡

Category	Reaction Number	Collision	Rate Coefficient	Refs
	229	He + He (3^3S) ⇒ He_2^+ + e	$3.625 \times 10^{-13} \times 10^{-2}$	[1] ‡
	230	He + He (3^3P) ⇒ He_2^+ + e	$5.7263 \times 10^{-11} \times 10^{-2}$	[1] ‡
	231	He + He (3^3D) ⇒ He_2^+ + e	$1.62 \times 10^{-10} \times 10^{-2}$	[1] ‡
	232	He + He (4^3S) ⇒ He_2^+ + e	$7.6 \times 10^{-10} \times 10^{-2}$	[1] ‡
	233	He + He (4^1S) ⇒ He_2^+ + e	$6.8 \times 10^{-10} \times 10^{-2}$	[1] ‡
	234	He + He (4^3P) ⇒ He_2^+ + e	$8.7 \times 10^{-10} \times 10^{-2}$	[1] ‡
	235	He + He (4^3D) ⇒ He_2^+ + e	$9.75 \times 10^{-10} \times 10^{-2}$	[1] ‡
	236	He + He (4^1D) ⇒ He_2^+ + e	$1.626 \times 10^{-9} \times 10^{-2}$	[1] ‡
	237	He + He (4^3F) ⇒ He_2^+ + e	$9.1973 \times 10^{-10} \times 10^{-2}$	[1] ‡
	238	He + He (4^1F) ⇒ He_2^+ + e	$9.26 \times 10^{-10} \times 10^{-2}$	[1] ‡
	239	He + He (4^1P) ⇒ He_2^+ + e	$4.7 \times 10^{-10} \times 10^{-2}$	[1] ‡
	240	He + He (3^3D) ⇒ He (3^3P) + He	6.7×10^{-11}	[1]
	241	He + He (3^1P) ⇒ He (3^1D) + He	5.8×10^{-10}	[1]
	242	He + He (4^3P) ⇒ He (4^3S) + He	10^{-10}	[1]
	243	He + He (4^3D) ⇒ He (4^3P) + He	9.3×10^{-11}	[1]
	244	He + He (4^3D) ⇒ He (4^3S) + He	5.8×10^{-11}	[1]
	245	He + He (4^1D) ⇒ He (4^1S) + He	2.2×10^{-10}	[1]
	246	He + He (4^3F) ⇒ He (4^3D) + He	1.8×10^{-10}	[1]
	247	He + He (4^3F) ⇒ He (4^3P) + He	6.6×10^{-11}	[1]
	248	He + He (4^3F) ⇒ He (4^3S) + He	4.1×10^{-11}	[1]
	249	He + He (4^1F) ⇒ He (4^1D) + He	5.9×10^{-10}	[1]
	250	He + He (4^1F) ⇒ He (4^1S) + He	1.6×10^{-10}	[1]
	251	He + He (4^1P) ⇒ He (4^1F) + He	7.6×10^{-10}	[1]

Category	Reaction Number	Collision	Rate Coefficient	Refs
	252	He + He (4^1P) ⇒ He (4^1D) + He	7.4×10^{-10}	[1]
	253	He + He (4^1P) ⇒ He (4^1S) + He	3.6×10^{-10}	[1]
	254	He + He (3^3P) ⇒ He (3^3D) + He	8.1×10^{-11}	[1]
	255	He + He (3^1D) ⇒ He (3^1P) + He	3.3×10^{-10}	[1]
	256	He + He (4^3S) ⇒ He (4^3P) + He	1.7×10^{-10}	[1]
	257	He + He (4^3P) ⇒ He (4^3D) + He	1.4×10^{-10}	[1]
	258	He + He (4^3S) ⇒ He (4^3D) + He	1.5×10^{-10}	[1]
	259	He + He (4^1S) ⇒ He (4^1D) + He	8.2×10^{-10}	[1]
	260	He + He (4^3D) ⇒ He (4^3F) + He	2.5×10^{-10}	[1]
	261	He + He (4^3P) ⇒ He (4^3F) + He	1.3×10^{-10}	[1]
	262	He + He (4^3S) ⇒ He (4^3F) + He	1.5×10^{-10}	[1]
	263	He + He (4^1D) ⇒ He (4^1F) + He	8.3×10^{-10}	[1]
	264	He + He (4^1S) ⇒ He (4^1F) + He	8.3×10^{-11}	[1]
	265	He + He (4^1F) ⇒ He (4^1P) + He	3.2×10^{-10}	[1]
	266	He + He (4^1D) ⇒ He (4^1P) + He	4.3×10^{-10}	[1]
	267	He + He (4^1S) ⇒ He (4^1P) + He	7.8×10^{-10}	[1]
	268	2He + He (2^1P) ⇒ 3He	1.8×10^{-31}	[1]
	269	2He + He (2^1S) ⇒ 3He	1.3×10^{-33}	[1]
	270	2He + He^+ ⇒ He_2^+ + He	$1.4 \times 10^{-31} (T_g/300)^{-0.6}$	[1]
	271	He + He_2^+ ⇒ 2He + He^+	$1.4 \times 10^{-6} T_g^{-0.67} \exp(-28100/T_g)$	[1]

Category	Reaction Number	Collision	Rate Coefficient	Refs
	272	He (2^3P) ⇒ He (2^3S) + hν (1083 nm)	1.022×10^7	[1]
	273	He (2^1P) ⇒ He (2^1S) + hν (2058.7 nm)	1.976×10^6	[1]
	274	He (2^1P) ⇒ He + hν (58.43 nm)	1.799×10^9	[1]
	275	He (3^1S) ⇒ He (2^1P) + hν (728.14 nm)	1.81×10^7	[1]
	276	He (3^1P) ⇒ He (2^1S) + hν (501.57 nm)	1.338×10^7	[1]
	277	He (3^1P) ⇒ He + hν (53.7 nm)	5.66×10^8	[1]
	278	He (3^1D) ⇒ He (2^1P) + hν (667.82 nm)	6.38×10^7	[1]
	279	He (3^3S) ⇒ He (2^3P) + hν (706.52 nm)	1.54×10^7	[1]
	280	He (3^3P) ⇒ He (2^3S) + hν (388.86 nm)	9.478×10^6	[1]
	281	He (3^3D) ⇒ He (2^3P) + hν (587.56 nm)	7.06×10^7	[1]
	282	He (4^1S) ⇒ He (2^1P) + hν (504.77 nm)	6.55×10^6	[1]
	283	He (4^1S) ⇒ He (3^1P) + hν (211.37 nm)	4.59×10^6	[1]
	284	He (4^1P) ⇒ He (2^1S) + hν (396.47 nm)	7.17×10^6	[1]
	285	He (4^1P) ⇒ He + hν (52.22 nm)	2.46×10^8	[1]
	286	He (4^1D) ⇒ He (2^1P) + hν (492.19 nm)	2.02×10^7	[1]
	287	He (4^1D) ⇒ He (3^1P) + hν (190.89 nm)	7.11×10^6	[1]
	288	He (4^1F) ⇒ He (3^1D) + hν (186.97 nm)	1.38×10^7	[1]
	289	He (4^3S) ⇒ He (2^3P) + hν (471.32 nm)	5.89×10^6	[1]

Category	Reaction Number	Collision	Rate Coefficient	Refs
	290	He (4^3S) ⇒ He (3^3P) + hν (211.25 nm)	3.62×10^6	[1]
	291	He (4^3P) ⇒ He (2^3S) + hν (318.77 nm)	5.05×10^6	[1]
	292	He (4^3D) ⇒ He (2^3P) + hν (447.15 nm)	2.51×10^7	[1]
	293	He (4^3D) ⇒ He (3^3P) + hν (170.02 nm)	6.68×10^6	[1]
	294	He (4^3F) ⇒ He (3^3D) + hν (186.85 nm)	1.55×10^6	[1]
C	295	C + OH ⇒ CO + H	8.3×10^{-11}	NIST
	296	C + NO ⇒ CO + N	3.49×10^{-11} $(T_g/298)^{-0.82}$	NIST
	297	C + O_2 ⇒ CO + O	2.61×10^{-11}	NIST
	298	C (2s2p^3 $^3D^o$) ⇒ C + hν (156.1 nm)	1.2821×10^8	[11]
	299	C (2s2p^3 $^3P^o$) ⇒ C + hν (132.93 nm)	2.2222×10^8	[11]
	300	C (2s2p3s $^3P^o$) ⇒ C + hν (165.72 nm)	3.2258×10^8	[11]
	301	C (2s2p4s $^3P^o$) ⇒ C + hν (128.04 nm)	1.1364×10^8	[11]
	302	C (2s2p3d $^3P^o$) ⇒ C + hν (126.13 nm)	1.7857×10^8	[11]
	303	C (2s2p3d $^3D^o$) ⇒ C + hν (127.75 nm)	2.2727×10^8	[11]
	304	C (2s2p3s $^1P^o$) ⇒ C (2s2p^2 ^1D) + hν (193.09 nm)	3.4483×10^8	[11]
	305	C (2s2p4s $^1P^o$) ⇒ C (2s2p^2 ^1D) + hν (146.75 nm)	5.8824×10^7	[11]
	306	C (2s2p3d $^1P^o$) ⇒ C (2s2p^2 ^1S) + hν (175.19 nm)	1.1905×10^8	[11]
	307	C (2s2p3d $^1D^o$) ⇒ C (2s2p^2 ^1D) + hν (148.18 nm)	3.8462×10^7	[11]
	308	C (2s2p3d $^1F^o$) ⇒ C (2s2p^2 ^1D) + hν (146.33 nm)	1.8868×10^8	[11]

Category	Reaction Number	Collision	Rate Coefficient	Refs
N^+	309	$N^+ + O \Rightarrow O^+ + N$	10^{-12}	[3]
	310	$N^+ + O + M \Rightarrow NO^+ + M$	10^{-29}	[3]
	311	$N^+ + O^- \Rightarrow O + N$	$2 \times 10^{-7} (300/T_g)^{0.5}$	[3]
	312	$N^+ + OH \Rightarrow OH^+ + N$	3.4×10^{-10}	[3]
	313	$N^+ + OH \Rightarrow NO^+ + H$	3.4×10^{-10}	[3]
	314	$N^+ + N + M \Rightarrow N_2^+ + M$	10^{-29}	[3]
	315	$N^+ + H_2O \Rightarrow H_2O^+ + N$	1.19×10^{-9}	[3]
	316	$N^+ + H_2O \Rightarrow NO^+ + H_2$	2.1×10^{-10}	[3]
	317	$N^+ + NO \Rightarrow NO^+ + N$	4.72×10^{-10}	[3]
	318	$N^+ + NO \Rightarrow N_2^+ + O$	8.33×10^{-11}	[3]
	319	$N^+ + NO \Rightarrow O^+ + N_2$	10^{-12}	[3]
	320	$N^+ + NO^- \Rightarrow NO + N$	$2 \times 10^{-7} (300/T_g)^{0.5}$	[3]
	321	$N^+ + O_2 \Rightarrow NO^+ + O$	2.7×10^{-10}	[3]
	322	$N^+ + O_2 \Rightarrow O^+ + NO$	2.8×10^{-11}	[3]
	323	$N^+ + O_2 \Rightarrow O_2^+ + N$	3×10^{-10}	[3]
	324	$N^+ + O_2^- \Rightarrow O_2 + N$	$2 \times 10^{-7} (300/T_g)^{0.5}$	[3]
	325	$N^+ + O_3^- \Rightarrow O_3 + N$	$2 \times 10^{-7} (300/T_g)^{0.5}$	[3]
	326	$N^+ + N_2O \Rightarrow NO^+ + N_2$	5.5×10^{-10}	[3]
	327	$N^+ + N_2O^- \Rightarrow N_2O + N$	$2 \times 10^{-7} (300/T_g)^{0.5}$	[3]
	328	$N^+ + NO_2 \Rightarrow NO_2^+ + N$	3×10^{-10}	[3]
	329	$N^+ + NO_2 \Rightarrow NO^+ + NO$	5×10^{-10}	[3]
	330	$N^+ + NO_2^- \Rightarrow NO_2 + N$	$2 \times 10^{-7} (300/T_g)^{0.5}$	[3]
	331	$N^+ + O_3 \Rightarrow NO^+ + O_2$	5×10^{-10}	[3]
	332	$N^+ + NO_3^- \Rightarrow NO_3 + N$	$2 \times 10^{-7} (300/T_g)^{0.5}$	[3]
N	333	$N + O^+ + M \Rightarrow NO^+ + M$	10^{-29}	[3]
	334	$N + O + M \Rightarrow NO + M$	$6.3 \times 10^{-33} \exp(140/T_g)$	[3]
	335	$N + O^- \Rightarrow NO + e$	2.6×10^{-10}	[5]
	336	$N + OH \Rightarrow H + NO$	7.5×10^{-11}	[3]
	337	$N + H_2O^+ \Rightarrow NO^+ + H_2$	1.9×10^{-10}	[3]
	338	$2N + M \Rightarrow N_2 + M$	$8.3 \times 10^{-34} \exp(500/T_g)$	[3]
	339	$N + N_2^+ \Rightarrow N^+ + N_2$	10^{-12}	[3]
	340	$N + NO^+ + M \Rightarrow N_2O^+ + M$	$10^{-29} (300/T_g)$	[3]

Category	Reaction Number	Collision	Rate Coefficient	Refs
	341	$N + NO \Rightarrow N_2 + O$	$2.1 \times 10^{-11} \exp(100/T_g)$	[3]
	342	$N + O_2^+ \Rightarrow NO^+ + O$	1.5×10^{-10}	[3]
	343	$N + O_2 \Rightarrow NO + O$	$1.5 \times 10^{-11} \exp(-3600/T_g)$	[3]
	344	$N + O_2^- \Rightarrow NO_2 + e$	5×10^{-10}	[5]
	345	$N + HO_2 \Rightarrow NO + OH$	$1.7 \times 10^{-11} \exp(-1000/T_g)$	[3]
	346	$N + NO_2 \Rightarrow N_2O + O$	$5.8 \times 10^{-12} \exp(220/T_g)$	[3]
	347	$N + NO_2 \Rightarrow N_2 + 2O$	9.1×10^{-13}	[3]
	348	$N + NO_2 \Rightarrow 2NO$	6×10^{-13}	[3]
	349	$N + NO_2 \Rightarrow N_2 + O_2$	7×10^{-13}	[3]
	350	$N + NO_2^- \Rightarrow N_2 + O_2 + e$	10^{-12}	[3]
	351	$N + O_3 \Rightarrow NO + O_2$	5×10^{-16}	[3]
	352	$N + NO_3^- \Rightarrow N_2 + O_3 + e$	10^{-12}	[3]
O^+	353	$O^+ + O + M \Rightarrow O_2^+ + M$	10^{-29}	[3]
	354	$O^+ + O^- \Rightarrow 2O$	$2 \times 10^{-7} (300/T_g)^{0.5}$	[5]
	355	$O^+ + OH \Rightarrow O_2^+ + H$	3.6×10^{-10}	[3]
	356	$O^+ + OH \Rightarrow OH^+ + O$	3.3×10^{-10}	[3]
	357	$O^+ + OH^- \Rightarrow O + OH$	$2 \times 10^{-7} (300/T_g)^{0.5}$	[3]
	358	$O^+ + NO \Rightarrow NO^+ + O$	10^{-12}	[3]
	359	$O^+ + NO \Rightarrow O_2^+ + N$	3×10^{-12}	[3]
	360	$O^+ + H_2O \Rightarrow H_2O^+ + O$	3.2×10^{-9}	[3]
	361	$O^+ + N_2 + M \Rightarrow NO^+ + N + M$	$6 \times 10^{-29} (300/T_g)^2$	[3]
	362	$O^+ + NO^- \Rightarrow NO + O$	$2 \times 10^{-7} (300/T_g)^{0.5}$	[3]
	363	$O^+ + O_2 \Rightarrow O_2^+ + O$	$2.1 \times 10^{-11} (300/T_g)^{0.5}$	[3]
	364	$O^+ + O_2^- \Rightarrow O_2 + O$	$2 \times 10^{-7} (300/T_g)^{0.5}$	[3]
	365	$O^+ + N_2O \Rightarrow N_2O^+ + O$	6.3×10^{-10}	[3]
	366	$O^+ + N_2O \Rightarrow NO^+ + NO$	2.3×10^{-10}	[3]
	367	$O^+ + N_2O \Rightarrow O_2^+ + N_2$	2×10^{-11}	[3]
	368	$O^+ + N_2O^- \Rightarrow N_2O + O$	$2 \times 10^{-7} (300/T_g)^{0.5}$	[3]
	369	$O^+ + NO_2 \Rightarrow NO^+ + O_2$	5×10^{-10}	[3]
	370	$O^+ + NO_2 \Rightarrow NO_2^+ + O$	1.6×10^{-9}	[3]
	371	$O^+ + NO_2^- \Rightarrow NO_2 + O$	$2 \times 10^{-7} (300/T_g)^{0.5}$	[3]
	372	$O^+ + O_3 \Rightarrow O_2^+ + O_2$	10^{-10}	[3]

Category	Reaction Number	Collision	Rate Coefficient	Refs
	373	$O^+ + O_3^- \Rightarrow O_3 + O$	$2 \times 10^{-7} (300/T_g)^{0.5}$	[3]
	374	$O^+ + NO_3^- \Rightarrow NO_3 + O$	$2 \times 10^{-7} (300/T_g)^{0.5}$	[3]
O	375	$O + O^- \Rightarrow O_2 + e$	1.4×10^{-10}	[3]
	376	$O + OH \Rightarrow H + O_2$	$2.2 \times 10^{-11} \exp(-350/T_g)$	[3]
	377	$O + OH^- \Rightarrow HO_2 + e$	2×10^{-10}	[3]
	378	$O + H_2O^+ \Rightarrow O_2^+ + H_2$	5.5×10^{-11}	[3]
	379	$O + CO \Rightarrow CO_2$	$1.7 \times 10^{-33} \exp(-1509.5/T_g)$	NIST
	380	$O + N_2^+ \Rightarrow O^+ + N_2$	$1 \times 10^{-11} (300/T_g)^{0.5}$	[3]
	381	$O + N_2^+ \Rightarrow NO^+ + N$	1.4×10^{-10}	[3]
	382	$O + NO + M \Rightarrow NO_2 + M$	$10^{-31} (300/T_g)^{1.6}$	[3]
	383	$O + NO^- \Rightarrow O^- + NO$	3×10^{-10}	[3]
	384	$O + HNO \Rightarrow OH + NO$	5.99×10^{-11}	[3]
	385	$O + O_2 + M \Rightarrow O_3 + M$	$3.4 \times 10^{-34} (T_g/300)^{-1.2}$	[12]
	386	$O + O_2^- \Rightarrow O^- + O_2$	3.3×10^{-10}	[3]
	387	$O + O_2^- \Rightarrow O_3 + e$	1.5×10^{-10}	[3]
	388	$2O + M \Rightarrow O_2 + M$	$3.2 \times 10^{-35} \exp(900/T_g)$	[3]
	389	$O + HO_2 \Rightarrow OH + O_2$	$8.3 \times 10^{-11} \exp(-500/T_g)$	[3]
	390	$O + H_2O_2 \Rightarrow OH + HO_2$	$3.3 \times 10^{-11} \exp(-2950/T_g)$	[3]
	391	$O + CO_2 \Rightarrow CO + O_2$	2.81×10^{-11} $\exp(-26461/T_g)$	NIST
	392	$O + NO_2 \Rightarrow NO + O_2$	$6.5 \times 10^{-12} \exp(120/T_g)$	[3]
	393	$O + NO_2 + M \Rightarrow NO_3 + M$	$9 \times 10^{-32} (300/T_g)^2$	[3]
	394	$O + NO_2^- \Rightarrow NO_3 + e$	10^{-12}	[3]
	395	$O + HNO_2 \Rightarrow NO_2 + OH$	$2 \times 10^{-11} \exp(-3000/T_g)$	[3]
	396	$O + O_3 \Rightarrow 2O_2$	$8 \times 10^{-12} \exp(-2060/T_g)$	[3]
	397	$O + O_3^- \Rightarrow 2O_2 + e$	3×10^{-10}	[3]
	398	$O + O_3^- \Rightarrow O_2^- + O_2$	10^{-11}	[3]
	399	$O + NO_3 \Rightarrow O_2 + NO_2$	1.7×10^{-11}	[3]
	400	$O + NO_3^- \Rightarrow NO_2 + O_2 + e$	10^{-12}	[3]
O^-	401	$O^- + OH^+ \Rightarrow 2O + H$	10^{-7}	[3]
	402	$O^- + OH^+ \Rightarrow O + OH$	$2 \times 10^{-7} (300/T_g)^{0.5}$	[3]
	403	$O^- + N_2 \Rightarrow N_2O + e$	10^{-12}	[3]

Category	Reaction Number	Collision	Rate Coefficient	Refs
	404	$O^- + NO \Rightarrow NO_2 + e$	2.6×10^{-10}	[3]
	405	$O^- + NO + M \Rightarrow NO_2^- + M$	10^{-29}	[3]
	406	$O^- + N_2^+ \Rightarrow O + N_2$	$2 \times 10^{-7}\ (300/T_g)^{0.5}$	[3]
	407	$O^- + N_2^+ \Rightarrow O + 2N$	10^{-7}	[3]
	408	$O^- + NO^+ \Rightarrow 2O + N$	10^{-7}	[3]
	409	$O^- + NO^+ \Rightarrow O + NO$	$2 \times 10^{-7}\ (300/T_g)^{0.5}$	[3]
	410	$O^- + O_2^+ \Rightarrow 3O$	10^{-7}	[3]
	411	$O^- + O_2^+ \Rightarrow O + O_2$	$2 \times 10^{-7}\ (300/T_g)^{0.5}$	[3]
	412	$O^- + O_2 \Rightarrow O_3 + e$	10^{-12}	[3]
	413	$O^- + O_2 + M \Rightarrow O_3^- + M$	$1.1 \times 10^{-30}\ (300/T_g)$	[3]
	414	$O^- + H_2O^+ \Rightarrow O + OH + H$	10^{-7}	[3]
	415	$O^- + H_2O^+ \Rightarrow O + H_2O$	$2 \times 10^{-7}\ (300/T_g)^{0.5}$	[3]
	416	$O^- + H_2O \Rightarrow OH^- + OH$	1.4×10^{-9}	[3]
	417	$O^- + N_2O^+ \Rightarrow O + N_2O$	$2 \times 10^{-7}\ (300/T_g)^{0.5}$	[3]
	418	$O^- + N_2O^+ \Rightarrow 2O + N_2$	10^{-7}	[3]
	419	$O^- + N_2O \Rightarrow NO^- + NO$	2×10^{-10}	[3]
	420	$O^- + N_2O \Rightarrow N_2O^- + O$	2×10^{-12}	[3]
	421	$O^- + NO_2^+ \Rightarrow O + NO_2$	$2 \times 10^{-7}\ (300/T_g)^{0.5}$	[3]
	422	$O^- + NO_2^+ \Rightarrow O + N + O_2$	10^{-7}	[3]
	423	$O^- + NO_2 \Rightarrow NO_2^- + O$	1.2×10^{-9}	[3]
	424	$O^- + O_3 \Rightarrow O_3^- + O$	8×10^{-10}	[3]
	425	$O^- + O_3 \Rightarrow 2O_2 + e$	3×10^{-10}	[3]
	426	$O^- + NO_3 \Rightarrow NO_3^- + O$	3×10^{-10}	[3]
OH^+	427	$OH^+ + OH \Rightarrow H_2O^+ + O$	7×10^{-10}	[3]
	428	$OH^+ + OH^- \Rightarrow 2OH$	$2 \times 10^{-7}\ (300/T_g)^{0.5}$	[3]
	429	$OH^+ + OH^- \Rightarrow OH + O + H$	10^{-7}	[3]
	430	$OH^+ + H_2O \Rightarrow H_2O^+ + OH$	1.59×10^{-9}	[3]
	431	$OH^+ + NO \Rightarrow NO^+ + OH$	5.2×10^{-10}	[3]
	432	$OH^+ + NO^- \Rightarrow NO + OH$	$2 \times 10^{-7}\ (300/T_g)^{0.5}$	[3]
	433	$OH^+ + NO^- \Rightarrow NO + O + H$	10^{-7}	[3]
	434	$OH^+ + O_2 \Rightarrow O_2^+ + OH$	5.9×10^{-10}	[3]
	435	$OH^+ + O_2^- \Rightarrow O_2 + OH$	$2 \times 10^{-7}\ (300/T_g)^{0.5}$	[3]

Category	Reaction Number	Collision	Rate Coefficient	Refs
	436	$OH^+ + O_2^- \Rightarrow O_2 + O + H$	10^{-7}	[3]
	437	$OH^+ + N_2O \Rightarrow N_2O^+ + OH$	2.13×10^{-10}	[3]
	438	$OH^+ + N_2O^- \Rightarrow N_2O + OH$	$2 \times 10^{-7} (300/T_g)^{0.5}$	[3]
	439	$OH^+ + N_2O^- \Rightarrow N_2O + O + H$	10^{-7}	[3]
	440	$OH^+ + NO_2 \Rightarrow NO^+ + HO_2$	1.3×10^{-9}	[3]
	441	$OH^+ + NO_2^- \Rightarrow NO_2 + OH$	$2 \times 10^{-7} (300/T_g)^{0.5}$	[3]
	442	$OH^+ + NO_2^- \Rightarrow NO_2 + O + H$	10^{-7}	[3]
	443	$OH^+ + O_3^- \Rightarrow O_3 + OH$	$2 \times 10^{-7} (300/T_g)^{0.5}$	[3]
	444	$OH^+ + O_3^- \Rightarrow O_3 + O + H$	10^{-7}	[3]
	445	$OH^+ + NO_3^- \Rightarrow NO_3 + OH$	$2 \times 10^{-7} (300/T_g)^{0.5}$	[3]
	446	$OH^+ + NO_3^- \Rightarrow NO_3 + O + H$	10^{-7}	[3]
OH	447	$OH (A^2\Sigma^+ v0) \Rightarrow OH + h\nu$ (308.9 nm)	3.48×10^4	[13]
	448	$2OH \Rightarrow O + H_2O$	$8.8 \times 10^{-12} exp(-503/T_g)$	[3]
	449	$2OH + M \Rightarrow H_2O_2 + M$	$6.9 \times 10^{-31} (T_g/300)^{-0.8}$	[3]
	450	$OH + CO \Rightarrow CO_2 + H$	$5 \times 10^{-14} (T_g/298)^{1.5}$ exp $(250.1804/T_g)$	NIST
	451	$OH + NO + M \Rightarrow HNO_2 + M$	$7.4 \times 10^{-31} (300/T_g)^{2.4}$	[3]
	452	$OH + HNO \Rightarrow NO + H_2O$	$8 \times 10^{-11} exp(-500/T_g)$	[3]
	453	$OH + HO_2 \Rightarrow O_2 + H_2O$	$4.8 \times 10^{-11} exp(250/T_g)$	[3]
	454	$OH + H_2O_2 \Rightarrow HO_2 + H_2O$	$2.9 \times 10^{-12} exp(-160/T_g)$	[3]
	455	$OH + NO_2 + M \Rightarrow HNO3 + M$	$2.2 \times 10^{-30} (300/T_g)^{2.9}$	[3]
	456	$OH + HNO_2 \Rightarrow NO_2 + H_2O$	$1.8 \times 10^{-11} exp(-390/T_g)$	[3]
	457	$OH + O_3 \Rightarrow HO_2 + O_2$	$1.6 \times 10^{-12} exp(-1000/T_g)$	[3]
	458	$OH + NO_3 \Rightarrow HO_2 + NO_2$	2×10^{-11}	[3]
	459	$OH + HNO_3 \Rightarrow NO_3 + H_2O$	$1.5 \times 10^{-14} exp(650/T_g)$	[3]
OH^-	460	$OH^- + N^+ \Rightarrow N + OH$	$2 \times 10^{-7} (300/T_g)^{0.5}$	[3]
	461	$OH^- + H_2O^+ \Rightarrow OH + H_2O$	$2 \times 10^{-7} (300/T_g)^{0.5}$	[3]
	462	$OH^- + H_2O^+ \Rightarrow 2OH + H$	10^{-7}	[3]
	463	$OH^- + N_2^+ \Rightarrow OH + N_2$	$2 \times 10^{-7} (300/T_g)^{0.5}$	[3]
	464	$OH^- + N_2^+ \Rightarrow OH + 2N$	10^{-7}	[3]
	465	$OH^- + NO^+ \Rightarrow OH + NO$	$2 \times 10^{-7} (300/T_g)^{0.5}$	[3]

Category	Reaction Number	Collision	Rate Coefficient	Refs
	466	$OH^- + NO^+ \Rightarrow OH + N + O$	10^{-7}	[3]
	467	$OH^- + O_2^+ \Rightarrow OH + O_2$	$2 \times 10^{-7} (300/T_g)^{0.5}$	[3]
	468	$OH^- + O_2^+ \Rightarrow OH + 2O$	10^{-7}	[3]
	469	$OH^- + N_2O^+ \Rightarrow OH + N_2O$	$2 \times 10^{-7} (300/T_g)^{0.5}$	[3]
	470	$OH^- + N_2O^+ \Rightarrow OH + N_2 + O$	10^{-7}	[3]
	471	$OH^- + NO_2^+ \Rightarrow OH + NO_2$	$2 \times 10^{-7} (300/T_g)^{0.5}$	[3]
	472	$OH^- + NO_2^+ \Rightarrow OH + N + O_2$	10^{-7}	[3]
	473	$OH^- + NO_2 \Rightarrow NO_2^- + OH$	1.9×10^{-9}	[3]
	474	$OH^- + O_3 \Rightarrow O_3^- + OH$	9×10^{-10}	[3]
H_2O^+	475	$H_2O^+ + O_2 \Rightarrow O_2^+ + H_2O$	4.3×10^{-10}	[3]
	476	$H_2O^+ + O_2^- \Rightarrow O_2 + H_2O$	$2 \times 10^{-7} (300/T_g)^{0.5}$	[3]
	477	$H_2O^+ + O_2^- \Rightarrow O_2 + OH + H$	10^{-7}	[3]
	478	$H_2O^+ + NO \Rightarrow NO^+ + H_2O$	4.6×10^{-10}	[3]
	479	$H_2O^+ + NO^- \Rightarrow NO + H_2O$	$2 \times 10^{-7} (300/T_g)^{0.5}$	[3]
	480	$H_2O^+ + NO^- \Rightarrow NO + OH + H$	10^{-7}	[3]
	481	$H_2O^+ + N_2O^- \Rightarrow N_2O + H_2O$	$2 \times 10^{-7} (300/T_g)^{0.5}$	[3]
	482	$H_2O^+ + N_2O^- \Rightarrow N_2O + OH + H$	10^{-7}	[3]
	483	$H_2O^+ + NO_2 \Rightarrow NO_2^+ + H_2O$	1.2×10^{-9}	[3]
	484	$H_2O^+ + NO_2^- \Rightarrow NO_2 + H_2O$	$2 \times 10^{-7} (300/T_g)^{0.5}$	[3]
	485	$H_2O^+ + NO_2^- \Rightarrow NO_2 + OH + H$	10^{-7}	[3]
	486	$H_2O^+ + O_3^- \Rightarrow O_3 + H_2O$	$2 \times 10^{-7} (300/T_g)^{0.5}$	[3]
	487	$H_2O^+ + O_3^- \Rightarrow O_3 + OH + H$	10^{-7}	[3]
	488	$H_2O^+ + NO_3^- \Rightarrow NO_3 + H_2O$	$2 \times 10^{-7} (300/T_g)^{0.5}$	[3]
	489	$H_2O^+ + NO_3^- \Rightarrow NO_3 + OH + H$	10^{-7}	[3]
H_2O	490	$H_2O + N_2^+ \Rightarrow H_2O^+ + N_2$	2.3×10^{-9}	[3]
CO	491	$CO + O_2 \Rightarrow CO_2 + O$	$4.2 \times 10^{-12} \exp(-24056/T_g)$	NIST
	492	$CO + NO_2 \Rightarrow CO_2 + NO$	$1.16 \times 10^{-14} (T_g/298)^{3.44} \exp(-11425/T_g)$	NIST
	493	$CO + O_3 \Rightarrow CO_2 + O_2$	4×10^{-25}	NIST
	494	CO ($B^1\Sigma^+ v0$) $\Rightarrow$ CO ($A^1\Pi v3$) + hν (561.02 nm) $\Rightarrow$ CO (v10) + hν (204.63 nm)	2.9851×10^7	[14]

Category	Reaction Number	Collision	Rate Coefficient	Refs
	495	CO ($C^1\Sigma^+v0$) ⇒ CO ($A^1\Pi v3$) + hν (438.03 nm) ⇒ CO (v10) + hν (204.63 nm)	6.8027×10^8	[14]
	496	CO ($E^1\Pi v0$) ⇒ CO	1.5432×10^8	[14]†
N_2^+	497	$N_2^+ + NO \Rightarrow NO^+ + N_2$	3.9×10^{-10}	[3]
	498	$N_2^+ + NO^- \Rightarrow NO + N_2$	$2 \times 10^{-7} (300/T_g)^{0.5}$	[3]
	499	$N_2^+ + NO^- \Rightarrow NO + 2N$	10^{-7}	[3]
	500	$N_2^+ + O_2 \Rightarrow O_2^+ + N_2$	5×10^{-11}	[3]
	501	$N_2^+ + O_2^- \Rightarrow O_2 + N_2$	$2 \times 10^{-7} (300/T_g)^{0.5}$	[3]
	502	$N_2^+ + O_2^- \Rightarrow O_2 + 2N$	10^{-7}	[3]
	503	$N_2^+ + N_2O \Rightarrow N_2O^+ + N_2$	6×10^{-10}	[3]
	504	$N_2^+ + N_2O \Rightarrow NO^+ + N + N_2$	4×10^{-10}	[3]
	505	$N_2^+ + N_2O^- \Rightarrow N_2O + N_2$	$2 \times 10^{-7} (300/T_g)^{0.5}$	[3]
	506	$N_2^+ + N_2O^- \Rightarrow N_2O + 2N$	10^{-7}	[3]
	507	$N_2^+ + NO_2 \Rightarrow NO^+ + N_2O$	5×10^{-11}	[3]
	508	$N_2^+ + NO_2 \Rightarrow NO_2^+ + N_2$	3×10^{-10}	[3]
	509	$N_2^+ + NO_2^- \Rightarrow NO_2 + N_2$	$2 \times 10^{-7} (300/T_g)^{0.5}$	[3]
	510	$N_2^+ + NO_2^- \Rightarrow NO_2 + 2N$	10^{-7}	[3]
	511	$N_2^+ + O_3^- \Rightarrow O_3 + N_2$	$2 \times 10^{-7} (300/T_g)^{0.5}$	[3]
	512	$N_2^+ + O_3^- \Rightarrow O_3 + 2N$	10^{-7}	[3]
	513	$N_2^+ + O_3 \Rightarrow O_2^+ + O + N_2$	10^{-10}	[3]
	514	$N_2^+ + NO_3^- \Rightarrow NO_3 + N_2$	$2 \times 10^{-7} (300/T_g)^{0.5}$	[3]
	515	$N_2^+ + NO_3^- \Rightarrow NO_3 + 2N$	10^{-7}	[3]
N_2	516	N_2^+ ($B^2\Sigma_u^+v0$) ⇒ N_2^+ + hν (391.44 nm)	6.53×10^8	[15]
	517	N_2^+ ($B^2\Sigma_u^+v0$) ⇒ N_2^+ ($X^2\Sigma_g^+v1$) + hν (427.81 nm)	6.34×10^8	[15]
	518	$N_2 + O_2^+ \Rightarrow NO^+ + NO$	10^{-17}	[3]
	519	$N_2 + O_2^- \Rightarrow N_2 + O_2 + e$	$1.9 \times 10^{-12} (T_g/300)^{0.5} \exp(-4990/T_g)$	[3]
	520	N_2^+ ($B^2\Sigma_u^+v0$) + N_2 ⇒ $N_2^+ + N_2$	4.2×10^{-9}	[16]
	521	N_2 ($C^3\Pi_u v0$) ⇒ N_2 ($B^3\Pi_g v0$) + hν (337.13 nm)	12.1×10^6	[17]

Category	Reaction Number	Collision	Rate Coefficient	Refs
	522	$N_2\,(C^3\Pi_u v0) \Rightarrow N_2\,(B^3\Pi_g v1)$ + hν (357.69 nm)	8.6×10^6	[17]
	523	$N_2\,(C^3\Pi_u v0) \Rightarrow N_2\,(B^3\Pi_g v2)$ + hν (380.49 nm)	3×10^6	[17]
	524	$N_2\,(C^3\Pi_u v0) \Rightarrow N_2\,(B^3\Pi_g v3)$ + hν (405.94 nm)	10^6	[17]
	525	$N_2\,(C^3\Pi_u v1) \Rightarrow N_2\,(B^3\Pi_g v0)$ + hν (315.93 nm)	11×10^6	[17]
	526	$N_2\,(C^3\Pi_u v1) \Rightarrow N_2\,(B^3\Pi_g v2)$ + hν (353.67 nm)	5.2×10^6	[17]
	527	$N_2\,(C^3\Pi_u v1) \Rightarrow N_2\,(B^3\Pi_g v3)$ + hν (375.54 nm)	4.6×10^6	[17]
	528	$N_2\,(C^3\Pi_u v1) \Rightarrow N_2\,(B^3\Pi_g v4)$ + hν (399.85 nm)	2.4×10^6	[17]
	529	$N_2\,(C^3\Pi_u v2) \Rightarrow N_2\,(B^3\Pi_g v4)$ + hν (371.05 nm)	3.12×10^6	[17]
NO^+	530	$NO^+ + NO^- \Rightarrow 2NO$	$2 \times 10^{-7}\,(300/T_g)^{0.5}$	[3]
	531	$NO^+ + NO^- \Rightarrow NO + N + O$	10^{-7}	[3]
	532	$NO^+ + O_2^- \Rightarrow O_2 + N + O$	10^{-7}	[3]
	533	$NO^+ + O_2^- \Rightarrow O_2 + NO$	$2 \times 10^{-7}\,(300/T_g)^{0.5}$	[3]
	534	$NO^+ + N_2O^- \Rightarrow N_2O + NO$	$2 \times 10^{-7}\,(300/T_g)^{0.5}$	[3]
	535	$NO^+ + N_2O^- \Rightarrow N_2O + N + O$	10^{-7}	[3]
	536	$NO^+ + NO_2^- \Rightarrow NO_2 + NO$	$2 \times 10^{-7}\,(300/T_g)^{0.5}$	[3]
	537	$NO^+ + NO_2^- \Rightarrow NO_2 + N + O$	10^{-7}	[3]
	538	$NO^+ + O_3 \Rightarrow NO_2^+ + O_2$	10^{-15}	[3]
	539	$NO^+ + O_3^- \Rightarrow O_3 + N + O$	10^{-7}	[3]
	540	$NO^+ + O_3^- \Rightarrow O_3 + NO$	$2 \times 10^{-7}\,(300/T_g)^{0.5}$	[3]
	541	$NO^+ + NO_3^- \Rightarrow NO_3 + NO$	$2 \times 10^{-7}\,(300/T_g)^{0.5}$	[3]
	542	$NO^+ + NO_3^- \Rightarrow NO_3 + N + O$	10^{-7}	[3]
	543	$NO^+ + N_2O_5 \Rightarrow NO_2^+ + 2NO_2$	5.9×10^{-10}	[3]
NO	544	$NO + NO^- \Rightarrow 2NO + e$	5×10^{-12}	[3]
	545	$NO + O_2^+ \Rightarrow NO^+ + O_2$	4.6×10^{-10}	[3]
	546	$NO + HO_2 \Rightarrow OH + NO_2$	$3.4 \times 10^{-12}\,\exp(270/T_g)$	[3]

Category	Reaction Number	Collision	Rate Coefficient	Refs
	547	$NO + HO_2 \Rightarrow O_2 + HNO$	$3.3 \times 10^{-13} \exp(-1000/T_g)$	[3]
	548	$NO + CO_2 \Rightarrow CO + NO_2$	$1.57 \times 10^{-13} (T_g/298)^{3.15} \exp(-38610/T_g)$	NIST
	549	$NO + N_2O^+ \Rightarrow NO^+ + N_2O$	2.3×10^{-10}	[3]
	550	$NO + NO_2^+ \Rightarrow NO^+ + NO_2$	2.75×10^{-10}	[3]
	551	$NO + NO_2^- \Rightarrow NO^- + NO_2$	2.75×10^{-10}	[3]
	552	$NO + O_3 \Rightarrow NO_2 + O_2$	$1.8 \times 10^{-12} \exp(-1370/T_g)$	[3]
	553	$NO + O_3^- \Rightarrow NO_2^- + O_2$	10^{-11}	[3]
	554	$NO + O_3^- \Rightarrow NO_3^- + O$	10^{-11}	[3]
	555	$NO + NO_3 \Rightarrow 2NO_2$	$1.8 \times 10^{-11} \exp(110/T_g)$	[3]
	556	$NO + NO_3^- \Rightarrow NO_2^- + NO_2$	3×10^{-15}	[3]
NO^-	557	$NO^- + H_2 \Rightarrow NO + H_2 + e$	2.3×10^{-13}	[3]
	558	$NO^- + M \Rightarrow NO + M + e$	2.4×10^{-13}	[3]
	559	$NO^- + O_2^+ \Rightarrow NO + O_2$	$2 \times 10^{-7} (300/T_g)^{0.5}$	[3]
	560	$NO^- + O_2^+ \Rightarrow NO + 2O$	10^{-7}	[3]
	561	$NO^- + O_2 \Rightarrow O_2^- + NO$	5×10^{-10}	[3]
	562	$NO^- + NO_2^+ \Rightarrow NO + NO_2$	$2 \times 10^{-7} (300/T_g)^{0.5}$	[3]
	563	$NO^- + NO_2^+ \Rightarrow NO + N + O_2$	10^{-7}	[3]
	564	$NO^- + O_3 \Rightarrow O_3^- + NO$	3×10^{-10}	[3]
	565	$NO^- + N_2O^+ \Rightarrow NO + N_2O$	$2 \times 10^{-7} (300/T_g)^{0.5}$	[3]
	566	$NO^- + N_2O^+ \Rightarrow NO + N_2 + O$	10^{-7}	[3]
	567	$NO^- + N_2O \Rightarrow NO + N_2O + e$	5.1×10^{-12}	[3]
	568	$NO^- + N_2O \Rightarrow NO_2^- + N_2$	2.8×10^{-14}	[3]
	569	$NO^- + NO_2 \Rightarrow NO_2^- + NO$	3×10^{-10}	[3]
	570	$NO^- + NO_3 \Rightarrow NO_3^- + NO$	3×10^{-10}	[3]
O_2^+	571	$O_2^+ + O_2^- \Rightarrow 2O_2$	$2 \times 10^{-7} (300/T_g)^{0.5}$	[5]
	572	$O_2^+ + O_2^- \Rightarrow O_2 + 2O$	10^{-7}	[3]
	573	$O_2^+ + N_2O^- \Rightarrow N_2O + O_2$	$2 \times 10^{-7} (300/T_g)^{0.5}$	[3]
	574	$O_2^+ + N_2O^- \Rightarrow N_2O + 2O$	10^{-7}	[3]
	575	$O_2^+ + NO_2 \Rightarrow NO_2^+ + O_2$	6.6×10^{-10}	[3]
	576	$O_2^+ + NO_2 \Rightarrow NO^+ + O_3$	10^{-11}	[3]

Category	Reaction Number	Collision	Rate Coefficient	Refs
	577	$O_2^+ + NO_2^- \Rightarrow NO_2 + O_2$	$2 \times 10^{-7}\,(300/T_g)^{0.5}$	[3]
	578	$O_2^+ + NO_2^- \Rightarrow NO_2 + 2O$	10^{-7}	[3]
	579	$O_2^+ + O_3^- \Rightarrow O_3 + O_2$	$2 \times 10^{-7}\,(300/T_g)^{0.5}$	[3]
	580	$O_2^+ + O_3^- \Rightarrow O_3 + 2O$	10^{-7}	[3]
	581	$O_2^+ + NO_3^- \Rightarrow NO_3 + O_2$	$2 \times 10^{-7}\,(300/T_g)^{0.5}$	[3]
	582	$O_2^+ + NO_3^- \Rightarrow NO_3 + 2O$	10^{-7}	[3]
	583	$O_2^+ + N_2O_5 \Rightarrow NO_2^+ + NO_3 + O_2$	8.8×10^{-10}	[3]
O_2	584	$O_2 + O_2^- \Rightarrow 2O_2 + e$	$2.7 \times 10^{-10}(T_g/300)^{0.5}$ $\exp(-5590/T_g)$	[3]
	585	$O_2 + N_2O^+ \Rightarrow NO^+ + NO_2$	4.59×10^{-11}	[3]
	586	$O_2 + N_2O^+ \Rightarrow O_2^+ + N_2O$	2.24×10^{-10}	[3]
	587	$O_2 + HNO \Rightarrow NO + HO_2$	$5.25 \times 10^{-12}\exp(-1510/T_g)$	[3]
	588	$O_2 + HNO \Rightarrow NO_2 + OH$	1.66×10^{-15}	[3]
	589	$O_2 + O_3^- \Rightarrow O_3 + O_2 + e$	2.3×10^{-11}	[3]
HNO	590	$2HNO \Rightarrow N_2O + H_2O$	$1.4 \times 10^{-15}\exp(-1600/T_g)$	[3]
O_2^-	591	$O_2^- + O_3 \Rightarrow O_3 + O_2 + e$	6×10^{-10}	[3]
	592	$O_2^- + O_3 \Rightarrow O_3^- + O_2$	3.5×10^{-10}	[3]
	593	$O_2^- + N_2O^+ \Rightarrow O_2 + N_2O$	$2 \times 10^{-7}\,(300/T_g)^{0.5}$	[3]
	594	$O_2^- + N_2O^+ \Rightarrow O_2 + N_2 + O$	10^{-7}	[3]
	595	$O_2^- + N_2O \Rightarrow O_3^- + N_2$	10^{-11}	[3]
	596	$O_2^- + NO_2^+ \Rightarrow O_2 + NO_2$	$2 \times 10^{-7}\,(300/T_g)^{0.5}$	[3]
	597	$O_2^- + NO_2^+ \Rightarrow O_2 + N + O_2$	10^{-7}	[3]
	598	$O_2^- + NO_2 \Rightarrow NO_2^- + O_2$	7×10^{-10}	[3]
	599	$O_2^- + NO_3 \Rightarrow NO_3^- + O_2$	5×10^{-10}	[3]
	600	$O_2^- + HNO_3 \Rightarrow NO_3^- + HO_2$	2.8×10^{-10}	[3]
M	601	$M + O_3 \Rightarrow O + O_2 + M$	$3.92 \times 10^{-10}\exp$ $(-11400/T_g)$	[3]
HO_2	602	$2HO_2 \Rightarrow H_2O_2 + O_2$	$2.2 \times 10^{-13}\exp(600/T_g)$	[3]
	603	$HO_2 + O_3 \Rightarrow OH + 2O_2$	$1.4 \times 10^{-14}\exp(-600/T_g)$	[3]
	604	$HO_2 + NO_3 \Rightarrow NO_2 + OH + O_2$	4.8×10^{-12}	[3]
	605	$HO_2 + NO_3 \Rightarrow HNO_3 + O_2$	9.2×10^{-13}	[3]

Category	Reaction Number	Collision	Rate Coefficient	Refs
N_2O^+	606	$N_2O^+ + N_2O \Rightarrow NO^+ + NO + N_2$	1.2×10^{-11}	[3]
	607	$N_2O^+ + N_2O^- \Rightarrow 2N_2O$	$2 \times 10^{-7} (300/T_g)^{0.5}$	[3]
	608	$N_2O^+ + N_2O^- \Rightarrow N_2O + N_2 + O$	10^{-7}	[3]
	609	$N_2O^+ + NO_2 \Rightarrow NO^+ + N_2 + O_2$	4.29×10^{-10}	[3]
	610	$N_2O^+ + NO_2 \Rightarrow NO_2^+ + N_2O$	2.21×10^{-10}	[3]
	611	$N_2O^+ + NO_2^- \Rightarrow NO_2 + N_2O$	$2 \times 10^{-7} (300/T_g)^{0.5}$	[3]
	612	$N_2O^+ + NO_2^- \Rightarrow NO_2 + N_2 + O$	10^{-7}	[3]
	613	$N_2O^+ + O_3^- \Rightarrow O_3 + N_2O$	$2 \times 10^{-7} (300/T_g)^{0.5}$	[3]
	614	$N_2O^+ + O_3^- \Rightarrow O_3 + N_2 + O$	10^{-7}	[3]
	615	$N_2O^+ + NO_3^- \Rightarrow NO_3 + N_2O$	$2 \times 10^{-7} (300/T_g)^{0.5}$	[3]
	616	$N_2O^+ + NO_3^- \Rightarrow NO_3 + N_2 + O$	10^{-7}	[3]
N_2O	617	$N_2O + NO_2^- \Rightarrow NO_3^- + N_2$	5×10^{-13}	[3]
N_2O^-	618	$N_2O^- + NO_2^+ \Rightarrow N_2O + NO_2$	$2 \times 10^{-7} (300/T_g)^{0.5}$	[3]
	619	$N_2O^- + NO_2^+ \Rightarrow N_2O + N + O_2$	10^{-7}	[3]
NO_2^+	620	$NO_2^+ + NO_2^- \Rightarrow 2NO_2$	$2 \times 10^{-7} (300/T_g)^{0.5}$	[3]
	621	$NO_2^+ + NO_2^- \Rightarrow NO_2 + N + O_2$	10^{-7}	[3]
	622	$NO_2^+ + O_3^- \Rightarrow O_3 + NO_2$	$2 \times 10^{-7} (300/T_g)^{0.5}$	[3]
	623	$NO_2^+ + O_3^- \Rightarrow O_3 + N + O_2$	10^{-7}	[3]
	624	$NO_2^+ + NO_3^- \Rightarrow NO_3 + NO_2$	$2 \times 10^{-7} (300/T_g)^{0.5}$	[3]
	625	$NO_2^+ + NO_3^- \Rightarrow NO_3 + N + O_2$	10^{-7}	[3]
NO_2	626	$NO_2 + NO_2^- \Rightarrow NO_3^- + NO$	4×10^{-12}	[3]
	627	$NO_2 + O_3 \Rightarrow NO_3 + O_2$	$1.4 \times 10^{-13} \exp(-2470/T_g)$	[3]
	628	$NO_2 + O_3^- \Rightarrow NO_2^- + O_3$	7×10^{-11}	[3]
	629	$NO_2 + O_3^- \Rightarrow NO_3^- + O_2$	2×10^{-11}	[3]
	630	$NO_2 + NO_3 \Rightarrow NO_2 + NO + O_2$	$2.3 \times 10^{-13} \exp(-1600/T_g)$	[3]
NO_2^-	631	$NO_2^- + O_3 \Rightarrow NO_3^- + O_2$	1.8×10^{-11}	[3]
	632	$NO_2^- + NO_3 \Rightarrow NO_3^- + NO_2$	5×10^{-10}	[3]
	633	$NO_2^- + HNO_3 \Rightarrow NO_3^- + HNO_2$	1.6×10^{-9}	[3]
	634	$NO_2^- + N_2O_5 \Rightarrow$ $NO_3^- + NO_3 + NO$	7×10^{-10}	[3]
O_3	635	$O_3 + O_3^- \Rightarrow 3O_2 + e$	3×10^{-10}	[3]

Category	Reaction Number	Collision	Rate Coefficient	Refs
O_3^-	636	$O_3^- + NO_3 \Rightarrow NO_3^- + O_3$	5×10^{-10}	[3]
HNO_2	637	$2HNO_2 \Rightarrow NO + NO_2 + H_2O$	10^{-20}	[3]
	638	$HNO_2 + HNO_3 \Rightarrow 2NO_2 + H_2O$	1.6×10^{-17}	[3]
NO_3	639	$2NO_3 \Rightarrow 2NO_2 + O_2$	$5 \times 10^{-12} \exp(-3000/T_g)$	[3]
Crutzen Reactions with CH_4 and CH_3	640	$H + CH_3 \Rightarrow CH_2 + H_2$	$10^{-10} \exp[-63190/(R \times T_g)]$	NIST
	641	$H + CH_3 \Rightarrow CH_4$	3.5×10^{-10}	NIST
	642	$H + CH_4 \Rightarrow H_2 + CH_3$	$5.83 \times 10^{-13} (T_g/298)^3 \exp[-33590/(R \times T_g)]$	NIST
	643	$H_2 + CH_3 \Rightarrow CH_4 + H$	$6.86 \times 10^{-14} (T_g/298)^{2.74} \exp[-39410/(R \times T_g)]$	NIST
	644	$N + CH_3 \Rightarrow 2H + HCN$	3.32×10^{-13}	NIST
	645	$NH + CH_4 \Rightarrow CH_3 + NH_2$	$4.12 \times 10^{-12} (T_g/298)^{2.28} \exp[-85080/(R \times T_g)]$	NIST
	646	$O + CH_4 \Rightarrow OH + CH_3$	$2.26 \times 10^{-12} (T_g/298)^{2.2} \exp[-31760/(R \times T_g)]$	NIST
	647	$CH_2 + CH_4 \Rightarrow 2CH_3$	$5.44 \times 10^{-11} (T_g/298)^{1.63} \exp[-31930/(R \times T_g)]$	NIST
	648	$NH_2 + CH_4 \Rightarrow CH_3 + NH_3$	$8.77 \times 10^{-15} (T_g/298)^3 \exp[-17710/(R \times T_g)]$	NIST
	649	$OH + CH_4 \Rightarrow CH_3 + H_2O$	$4.16 \times 10^{-13} (T_g/298)^{2.18} \exp[-10240/(R \times T_g)]$	NIST
	650	$OH + CH_3 \Rightarrow H_2O + CH_2$	$1.2 \times 10^{-10} \exp[-11640/(R \times T_g)]$	NIST
	651	$OH + CH_3 \Rightarrow CH_4 + O$	$3.22 \times 10^{-14} (T_g/298)^{2.2} \exp[-18620/(R \times T_g)]$	NIST
	652	$2CH_3 \Rightarrow CH_4 + CH_2$	5.5×10^{-11}	NIST
	653	$CH_3 + H_2O_2 \Rightarrow CH_4 + HO_2$	$2.01 \times 10^{-14} \exp[2490/(R \times T_g)]$	NIST
	654	$CH_3 + HO_2 \Rightarrow CH_4 + O_2$	5.99×10^{-12}	NIST
	655	$CH_3 + H_2O \Rightarrow CH_4 + OH$	$1.64 \times 10^{-15} (T_g/298)^{3.31} \exp[-52550/(R \times T_g)]$	NIST

Category	Reaction Number	Collision	Rate Coefficient	Refs
	656	$CH_4 + NO_3 \Rightarrow CH_3 + HNO_3$	10^{-18}	NIST
	657	$CH_4 + NO_2 \Rightarrow CH_3 + HNO_2$	$1.26 \times 10^{-12} (T_g/298)^2$ $\exp[-218000/(R \times T_g)]$	NIST
	658	$CH_4 + HO_2 \Rightarrow CH_3 + H_2O_2$	3.01×10^{-13} $\exp[-77740/(R \times T_g)]$	NIST
	659	$CH_4 \Rightarrow H + CH_3$	$1.25 \times (T_g/298)^{-3.73}$ $\exp[-446000/(R \times T_g)]$	NIST
Expan-sion	660–1559	Photon Reactions	See Figure A.1	See Below

T_g and T_e are both in [K].

The rate coefficients of photon emissions are in [s^{-1}].

The rate coefficients of two–body collisions are in [cm^3/s], and the ones of three–body collisions are in [cm^6/s].

The species M represents He, N_2, and O_2 which plays a major part of the mixture.

*For T_e < 3.3 eV, the value is 0.

†Wavelength has not been found yet

‡Modified from the original reference to make the simulation with dt = 10^{-6}s fit the experimental observations

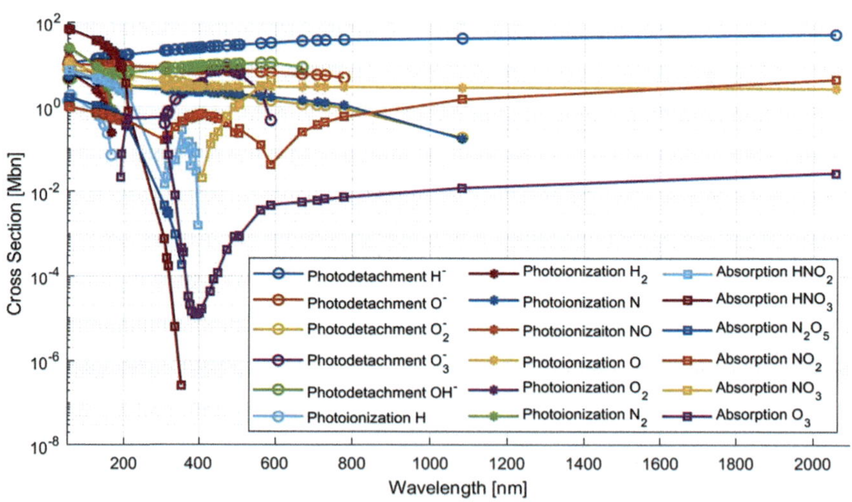

Figure A.1

References

1. T. Belmonte, R. P. Cardoso, G. Henrion, and F. Kosior, "Collisional – radiative modelling of a helium microwave plasma in a resonant," *J. Phys. D: Appl. Phys.*, no. 40, pp. 7343–7356, 2007, doi: 10.1088/0022-3727/40/23/015.
2. G. J. M. Hagelaar and L. C. Pitchford, "Solving the Boltzmann equation to obtain electron transport coefficients and rate coefficients for fluid models," *Plasma Sources Sci Technol*, vol. 14, no. 4, pp. 722–733, 2005, doi: 10.1088/0963-0252/14/4/011.
3. Y. Sakiyama, D. B. Graves, H. W. Chang, T. Shimizu, and G. E. Morfill, "Plasma chemistry model of surface microdischarge in humid air and dynamics of reactive neutral species," *J Phys D Appl Phys*, vol. 45, no. 42, 2012, doi: 10.1088/0022-3727/45/42/425201.
4. Sultana N. Nahar and A. K. Pradhan, "Electron–ion recombination rate coefficients, photoionization cross sections, and ionization fractions for astrophysically abundant elements. I. Carbon and nitrogen," *Astrophys J Suppl Ser*, vol. 111, pp. 339–355, 1997.
5. I. A. Kossyi, A. Y. Kostinsky, A. A. Matveyev, and V. P. Silakov, "Kinetic scheme of the non-equilibrium discharge in nitrogen-oxygen mixtures," *Plasma Sources Sci Technol*, vol. 1, no. 3, pp. 207–220, 1992, doi: 10.1088/0963-0252/1/3/011.
6. O. Nagy, "$X^2\Sigma_g^+$ ($v''=0$) $\rightarrow$ $B^2\Sigma_u^+$ ($v'=0$) excitation cross-sections of N_2^+ molecular ion by electron impact and the vibrational energy levels of the three target states N_2^+ ($X^2\Sigma_g^+$, $A^2\Pi_u$ and $B^2\Sigma_u^+$)," *Chem Phys*, vol. 286, pp. 109–114, 2003.
7. R. E. Center, "Electron–ion recombination measurements in CO at high pressures," *J Appl Phys*, vol. 44, no. 8, pp. 3538–3542, 1973, doi: 10.1063/1.1662797.
8. G. A. Germany, R. J. Anderson, and G. J. Salamo, "Electron impact excitation of the 3p (5P) state of atomic oxygen," *J Chem Phys*, vol. 89, p. 1999, 1988.
9. R. A. Gutcheck and E. C. Zipf, "Excitation of the CO fourth positive system by the dissociative recombination of CO_2^+ ions," *J Geophys Res*, vol. 78, no. 25, pp. 5429–5436, Sep. 1973, doi: 10.1029/ja078i025p05429.
10. R. Riahi, P. Teulet, Z. Ben Lakhdar, and A. Gleizes, "Cross-section and rate coefficient calculation for electron impact excitation, ionisation and dissociation of H_2 and OH molecules," *The European Physical Journal D*, vol. 40, pp. 223–230, 2006, doi: 10.1140/epjd/e2006-00159-2.

11. R. R. Haar, L. J. Curtis, T. J. Kvale, D. J. Beideck, I. Martinson, and R. Hellborg, "Lifetimes of excited levels in neutral carbon," *Astron Astrophys*, vol. 241, pp. 321–326, 1991.
12. S. A. Norberg, E. Johnsen, and M. J. Kushner, "Formation of reactive oxygen and nitrogen species by repetitive negatively pulsed helium atmospheric pressure plasma jets propagating into humid air," *Plasma Sources Sci Technol*, vol. 24, no. 3, p. 35026, 2015, doi: 10.1088/0963-0252/24/3/035026.
13. M. H. Stevens and R. R. Conway, "Calculated OH $A^2\Sigma^+–X^2\Pi$ (0,0) band rotational emission rate factors for solar resonance fluorescence," *J Geophys Res*, vol. 104, no. D13, pp. 16369–16378, 1999.
14. K. Kirby and D. L. Cooper, "Theoretical study of low-lying $1\Sigma^+$ and 1π states of CO. II. Transition dipole moments, oscillator strengths, and radiative lifetimes," *J Chem Phys*, vol. 90, no. 9, pp. 4895–4902, 1989, doi: 10.1063/1.456584.
15. R. G. Bennett and F. W. Dalby, "Experimental determination of the oscillator strength of the first negative bands of N_2^+," *J Chem Phys*, vol. 31, no. 2, pp. 434–441, 1959.
16. G. I. Mackay and R. E. March, "Collisional deactivation rates of N_2^+ $(B^2\Sigma_u^+)$," *Can J Chem*, vol. 49, p. 1268, 1970.
17. A. W. Johnson and R. G. Fowler, "Measured lifetimes of rotational and vibrational levels of electronic states of N_2," *J. Chem. Phys.*, vol. 65, no. 53, 1970.

Photoreaction References

Dissociations and Detachments

L. C. Lee and G. P. Smith, "Photodissociation and photodetachment of molecular negative ions. VI. Ions in $O_2/CH_4/H_2O$ mixtures from 3500 to 8600 Å," *J. Chem. Phys.*, vol. 70, p. 1727, 1979.

Ionizations

Geoffrey V. Marr, *Photoionization Processes in Gases*, Academic Press, London, 1967.

Other Absorptions

T. Moiseev, N. N. Misra, S. Patil, P. J. Cullen, P. Bourke, K. M. Keener, and J. P. Mosnier, "Post-discharge gas composition of a large-gap DBD in humid air by UV-Vis absorption spectroscopy," *Plasma Sources Sci. Technol.*, vol. 23, 065033, 2014.

Index

For Product Safety Concerns and Information please contact our EU representative GPSR@taylorandfrancis.com Taylor & Francis Verlag GmbH, Kaufingerstraße 24, 80331 München, Germany

Batch number: 10726509

Printed by Printforce, the Netherlands